Straight Talk

The No-Nonsense Guide to Strategic AI Adoption

Frameworks and strategies business leaders need to implement AI responsibly, integrate it with operations, and lead their teams with confidence.

Andrea M. Hill

Straight Talk: The No-Nonsense Guide to Strategic AI Adoption
Copyright © 2023, 2025 by Andrea M. Hill.
All rights reserved.

Published by Mardrea Press, Chicago, Illinois

MARDREA PRESS

First edition published by Mardrea Press in 2023
Updated edition published by Mardrea Press in 2025

Cover Design: Jenny Knuth. All rights reserved.
Developmental Editing: Raphael J. Hill, John O'Hara
Copy Editors: Sherri Burney, Karen Rice
Proofreaders: Kristin Kopaz, Elisabetta Marini, Kieran Stuck

Library of Congress Control Number: 2025920397
ISBN 978-0-9916001-1-3 (Paperback)
ISBN 978-0-9916001-6-8 (Hard Cover)
ISBN 978-0-9916001-2-0 (Kindle)
ISBN 978-0-9916001-3-7 (ePub)

Printed in the United States of America

Table of Contents

○ ○ ○

Dedication

Russell J. Hill
September 11, 1931 - August 5, 2025

It was early autumn in 2023, and I was hanging out with my dad and stepmom at their house in Des Moines, Iowa. My then 92-year-old dad, with his characteristic look of earnestness and intensity, leaned in to me and said, "I need you to explain Artificial Intelligence. I just don't understand it."

I wasn't surprised by his interest. This was the same guy who, in the 1960s, made sure our stereo equipment was top-of-the-line. An avid amateur photographer all his life, he always had the latest camera equipment. We had the first color TV in our neighborhood. He had the best CD collection of anyone I knew when most people were still barely transitioning from tape.

In his 60s, after years in criminal and civil law, his appointment to the Federal Judiciary in the bankruptcy courts marked his transition into business law. He called me then, and we began an ongoing conversation about business technology and enterprise resource planning (ERP): what technology was available, how it worked, and what it meant for the businesses that used it. When he was in his late 70s, headed to Africa for a service trip followed by a photo safari, he had me set up a blog so he could more readily post his observations and pictures. During COVID-19 he spent time digitizing his massive slide and photo collection.

That day in 2023, after discussing AI in some detail, he sat back and said, "What an interesting time to be alive."

What an understatement. From his family's 1931 Philco 70 cathedral radio to the democratization of large language models, my dad lived through the most dramatic century of technical innovation the world has ever seen. He slipped away from us in August of 2025, as the final proofs were being done on this book.

It **is** an interesting time to be alive, Dad. I am so grateful for the love of technology you gifted to me, and from now on, will always miss having you in my life to share it with.

o o o

Preface

Writing about AI is unlike writing about almost any other business subject, because the field changes daily. To keep up, I have had to study it every day, tracking not only the technology itself but also the flood of articles, research reports, policy papers, and commentary that shape our understanding of it. The body of knowledge is enormous, and it is evolving constantly and dramatically.

Many of the ideas you will encounter here are now shared across dozens of sources, making it difficult to attribute any single one as the "first" reference. For ease of reading, I have not included inline notations unless I am quoting directly or drawing from a specific, referenceable source. At the end of the book, you will find an extensive "Notes and Sources" appendix that lists the works I have relied on most heavily.

Introduction

When I first wrote this book in March 2023, I warned that it would be outdated within a year and a relic in 18 months. That prediction turned out to be correct. In just two years, artificial intelligence (AI) has gone from being a curiosity to an everyday business tool. AI is no longer something businesses debate whether or not to adopt, because it's already built into the software you use to run your company, whether that's your Customer Relationship Management (CRM), Enterprise Resource Management (ERP), email platform, marketing automation, or accounting system. Now the challenge for small and midsize businesses is how to use it responsibly and strategically, while also managing the risks you did not anticipate.

It is fair to acknowledge that this is a lot. For many business owners, the work of AI governance feels like one more set of responsibilities they never asked for. Yet opting out is no longer realistic. Just as you would never decide to skip customer service because it is difficult or expensive, you cannot skip the work of making sure AI is used responsibly. The world has changed, and this oversight is now simply part of doing business.

This guide is designed to make that work manageable. It will provide you with a working knowledge of AI, highlight the ethical challenges AI presents, and outline the legal considerations that come with using AI in your business. Perhaps most usefully, it will offer a framework for AI software application, planning, and implementation that helps you manage risks while taking advantage of AI's real opportunities to improve performance, efficiency, and growth.

AI Skills: What Really Matters

Headlines say managers are fixated on hiring employees that already have AI skills, but crafting ChatGPT prompts is the easy part. You can teach someone how to use AI tools in hours.

What's hard, and actually drives business value, is judgment, strategy, and leadership. AI can speed up drafts and ideas, but it can't provide nuance, context, or brand voice. That's still your job.

One thing will not change: AI will keep evolving faster than any book can capture. The principles of responsible and ethical adoption, including transparency, accountability, fairness, and alignment with your company's values, will hold up no matter how the technology develops. Think of this book as a compass rather than a map. It will not show you every twist in the road, but it will help you navigate wisely as the terrain continues to change.

AI is rewriting the rules of business in real time. The question is no longer whether to engage with AI, but how effectively you'll step in. This book is here to support you and push you forward.

The State of AI (2025)

Artificial intelligence has gone mainstream, but not in the way many expected. Two years ago, the story was all about breakthroughs, billion-dollar valuations, and the sense that we were entering a new industrial revolution. That narrative is still alive. Sam Altman, the CEO of OpenAI, recently admitted that we are "probably in a bubble," even as he continues to be one of the loudest champions of AI. That paradox captures where we are today: AI is both infrastructure and hype, both an everyday tool and an overblown promise.

Market Growth and Momentum

Analysts now estimate the global AI market in the hundreds of billions, with trillions of dollars in business value (new revenue, cost savings, and competitive advantage) expected before the decade is over. Venture funding has kept pace, flowing at a rate of tens of billions annually into AI platforms, startups, and chip makers.

Keep in mind that market size does not equal maturity. Many of the products and companies attracting attention today won't survive. We've seen this before, from the dot-coms in the early 2000s, to blockchain in the 2010s, and now with AI. The winners will be the firms that create lasting value, not those that simply ride the hype wave.

Adoption Among SMBs

In this book, I'll use "SMB" to mean small and midsize businesses. These are the companies that make up the vast majority of the economy, but rarely have the resources of the Fortune 500. While this book is applicable to businesses of any size, I've written it with SMBs in mind.

For small and midsize businesses, adoption is already widespread. Surveys in late 2024 showed more than 60% of SMBs using AI in at least one function, with marketing and sales leading the way. Adoption does not always equate to strategic use. Many business owners don't even realize that AI is already built into the systems they use, from Microsoft Outlook suggesting email text, to Quick-Books predicting cash flow, to Shopify recommending pricing strategies.

The irony is that while AI is everywhere, business expectations are still colliding with limitations. Owners hoping AI could take entire workloads off their plates are finding that these systems still hallucinate facts, misunderstand context, and require human oversight. For example, while it's tempting to use AI to churn out customer emails, poorly supervised AI-generated messaging often comes across as tone-deaf, inauthentic, or flat-out wrong. The value isn't in letting AI replace your voice, but in using it to accelerate drafts and ideas while you and your employees bring judgment, nuance, brand personality, and human touch.

From Hype to Everyday Use

ChatGPT made AI feel suddenly accessible in 2023. In 2025, the story has matured: AI is quietly embedded in the everyday tools we already rely on. Microsoft Copilot drafts documents, Keap suggests marketing content, HubSpot automates prospect follow-up, Canva generates ad variations, Odoo predicts inventory needs. AI has become invisible infrastructure.

Yet the standalone AI-tool marketplace is crowded with products making extravagant claims. "This AI will run your entire marketing department." "This AI will replace your data team." "This AI will solve strategy." At best, most of these are partial solutions. At worst, they're security risks. Employees experimenting with unapproved

AI tools have already triggered costly data leaks, exposing sensitive customer information and trade secrets.

This is where the bubble shows up most clearly. There are too many tools chasing the same problems, too many vendors overpromising, and too many buyers assuming AI is further along than it is. The technology is real and the potential is massive, but none of these promises are worth betting your business on.

Economic and Workforce Implications

Media headlines often suggest AI will wipe out entire categories of jobs. The reality is more nuanced. AI is transforming tasks, not eliminating whole professions. Marketing assistants can spend less time writing first drafts and more time refining strategy. Customer service agents can spend less time answering FAQs and more time resolving complex issues. Bookkeepers can spend less time categorizing transactions and more time advising on financial health.

Just because AI hasn't replaced the workforce doesn't mean disruption isn't real. SMBs face two pressing challenges:

1. Upskilling and trust: Employees must be trained to use AI responsibly, which requires investment in both technical training and judgment. If people don't

understand how to question AI outputs, they risk taking flawed results at face value.

2. Workforce design: Leaders must think about how to redesign roles and workflows so that people and AI complement each other. If AI is doing more of the repetitive work, how do employees continue to grow, develop, and add value?

This is where SMBs have a hidden advantage: smaller companies are often more agile. They can redesign workflows, adopt new tools, and retrain staff faster than large enterprises.

AI and Company Culture

AI doesn't just change what work gets done; it will profoundly change how people feel about their work and their workplaces. If employees see AI as reducing their responsibilities or leaving less room for judgment, learning, or creativity, they will resist it. That resistance only grows stronger when they believe AI could eliminate their jobs altogether. If instead they see it as a tool that helps them work more efficiently and take on more meaningful challenges, they will embrace it.

Culture is built on the daily decisions we make and how employees feel about them. Will you use AI to speed through customer questions while still giving people a way to connect with a human when they need it? Will you use AI to police how people spend their time, or to highlight opportunities that make their work more effective? As a tool for control, or as a tool for coaching and support? Every decision will signal what you value, and collectively will refine your culture.

For small and midsize businesses, culture can be (should be!) a competitive advantage. You may not be able to pay the highest salaries, but you can create a workplace where people feel trusted and supported. If you are not mindful about how you bring AI into the business, it will shape the culture for you, likely in ways you don't want. If you approach AI with intention, it can reinforce the culture you want: the one you've chosen, defined, and nurtured. Don't let AI decide what kind of company you are or want to be. That's your job.

The Small Business Technology Gap

There's another layer to this conversation that's not getting enough attention. Small and midsize businesses aren't just wrestling with AI adoption; they're still catching up on basic digital infrastructure. Too many companies are still running on outdated systems, under-investing in cybersecurity, or relying on spreadsheets where competitors are using integrated ERPs and CRMs.

That's what I call the technology gap, though it's ultimately a competitiveness gap. If you don't have the digital backbone in place, you cannot take advantage of AI responsibly or effectively. The longer this gap persists, the harder it becomes to catch up, because your competitors aren't standing still; they're building more strategic data capabilities, stronger supply chains, and making faster, better-informed decisions.

For SMBs, this isn't just about buying software; it's about making sure your business has the digital maturity to stay competitive. AI will amplify whatever foundation you already have. If your systems are strong, it can give you leverage. If they're weak, it will magnify the cracks.

Regulation and Governance

Perhaps the clearest change since the 2023 version of this book is that regulation has begun to catch up. The European Union Artificial Intelligence Act (EU AI Act: 2024) is the most comprehensive law in place, setting requirements for transparency, risk classification, and compliance. In the U.S., the federal government issued its first Executive Order on AI in 2023, and agencies have been layering on guidance ever since. State-level laws, particularly in California, Illinois, and New York, are beginning to affect hiring, data use, and consumer privacy.

For SMBs, this means AI compliance isn't optional. Even if you're not the direct target of a regulation, your vendors are. That means their compliance obligations cascade down to you in the form of contract language, audit requirements, and customer disclosures. For businesses that have never had to think about compliance in these terms before, this is new territory.

The Competitive Reality

AI is already part of your business, built into the software and tools you use every day. It's also part of your competitors' businesses, which means the competitive baseline for efficiency, personalization, and customer experience has already started shifting. The catch? Not all AI is useful AI. Investing in superficial solutions or allowing employees to use unvetted "shadow AI" is just as likely to drain budgets or expose sensitive data as it is to create value.

AI disruption isn't only about jobs. It's about competitiveness. Companies that learn to integrate AI effectively will move ahead in productivity, decision-making, and customer experience. Those that don't will find themselves up against businesses that are faster, more relevant, and more efficient. Once a business falls behind its competitors, it's hard to catch up.

AI adoption cannot be left up to chance, because the minimum standards necessary to compete are changing. Even if you have no desire to be out there on the AI leading edge, you're already competing against companies that are. The real risk isn't being replaced by machines;

o o o

it's being replaced by competitors who know how to use those machines better than you do. For small and midsize businesses, the real decision isn't whether to use AI, but how to use it in ways that make you stronger instead of leaving you behind.

What This Means for You

AI in 2025 is a paradox. It's both real and overhyped. It's already built into your daily workflows, yet still shaky in many of its standalone uses. The businesses that thrive won't be the ones chasing shiny objects; they'll be the ones that:

- Use AI where it's practical and proven, not just where it looks trendy.
- Train their teams to see AI as a tool, not a replacement for human judgment.
- Put clear policies, rules, and safeguards in place to manage risk.
- Keep investing in people, not just in technology.

This book gives you the tools to do exactly that.

AI Readiness Starts Here

Now is a good time to step back and assess whether your business is ready for AI at all. Many small and midsize companies are tempted to start with seemingly simple applications, but the most important work happens behind the scenes.

Ask yourself:

- Do you have reliable, accurate data? AI depends on clean, consistent information. If your customer, financial, or inventory records are a mess, the outputs will be too.

- Are your systems connected? If sales, operations, and finance are still running in silos, AI can't give you the full picture. Integration matters more than algorithms.

- Is your team digitally confident? Employees don't need to become data scientists, but they do need to feel comfortable with new tools and willing to learn.

- Do you have a clear business case? AI is not a strategy by itself. It should solve a specific problem or create measurable value.

Getting these basics right makes any later investment in AI more effective and less risky. Think of it as laying the foundation before you build. Without readiness, even the best AI tools will deliver disappointing results.

Take the AI Readiness Assessment here: **werx.me/AI-ready**

Understanding AI

No doubt you're aware of the ongoing social debate around AI. A surface-level view might leave you thinking the biggest risk is that AI will one day "wake up," weaponize itself, and wipe out humanity. The trouble with those kinds of doomsday scenarios is not that they're unimaginable, but that they feel so far beyond our influence that most people tune out. It's like reading headlines about climate change or earthquakes: the scale can make us feel powerless, so we set the worries aside and get on with our day. When we focus only on the most extreme possibilities, we miss the practical issues of AI that we can influence right now.

Basic knowledge of AI, both how it is created and how it can be used, will make it easier to understand and to participate in the social conversation around AI. That conversation needs all of our informed engagement. The risks to individual businesses of using AI range from as small as a single off-brand blog post, to as large as finding yourself unable to sell your business or at the wrong end of a lawsuit. The social risks range from an increase in misinformation to, yes, weaponization. Between *where you sit now* and *the end of the world* is an actionable list of AI issues you **can** directly influence in your day-to-day life.

If you are worried that AI is something that will be difficult to understand, stop. You don't need to understand how to write code or connect smart machines (or even install your Ring doorbell) to understand the core principles of AI.

Machine Learning

Let's start with the most fundamental aspect of AI. Machine learning (according to so many experts in AI that one can't even cite the original source) is a lot like training a dog to do a trick. At first, you show the dog what you want it to do. Then you give it a treat each time it gets the trick right.

Similarly, in machine learning, you show a computer program lots of examples of something (let's say pictures of dogs and names of dog breeds), and you tell the program which ones are dogs and which ones are not.

Just as a dog may make mistakes at first when learning a new trick, the computer will also make mistakes when learning to recognize dogs. Over time, using more examples and consistent feedback, the computer gets better at recognizing dogs, until it recognizes dogs correctly every time. The data that is fed to the AI platform to teach it is called, appropriately, "training data."

That is all that machine learning is: teaching a computer program to learn by showing it lots of examples and giving it lots of feedback.

Of course, this simple analogy only scratches the surface. In reality, businesses have been using machine learning in practical ways for many years. When your bank flags a suspicious transaction on your credit card, that's machine learning. When your ecommerce platform suggests "customers also bought," that's machine learning. When your email program recommends words to complete your sentence, that's machine learning.

For small and midsize businesses, machine learning is less about building models from scratch and more about recognizing that the tools you already use, such as CRMs, accounting software, and marketing platforms, already have machine learning built in. In other words, you are already using AI even if you never signed up for an AI product.

Just like a dog learns whatever behavior you reward, the computer learns whatever data it is fed. If the training data is incomplete, outdated, or biased, the program will reproduce those flaws faithfully. For SMBs this is not abstract. A flawed model could recommend the wrong customers to target, discriminate against an employment candidate, over-order the wrong inventory, or flag legitimate transactions as fraud. In other words, machine learning is powerful, but it is also relentlessly literal.

Datasets

In order to teach a machine to learn, you need a dataset. A dataset is simply a collection of data. In the example of teaching a computer to recognize dogs, the dataset could be a database of all the dog pictures, videos, breed names, and breed descriptions in the world.

Several different types of datasets are used in AI development:

- A Training Dataset is the data that is used to train a machine learning model.

- A Validation Dataset is a subset of the training dataset; for example, just the dog pictures, videos, and breed names of dogs in the terrier category. Validation datasets are used to evaluate the performance of the machine learning model during its training.

- Public Datasets are datasets that are publicly available for machine learning research and development. Public datasets can include vast amounts of copyrighted and private data, data that the original author or owner did not give permission to include in the dataset.

- Private Datasets are owned by a specific group or individual and are not available to the public.

Curious about which datasets ChatGPT was trained on? Unfortunately, the company does not disclose which datasets were used, other than the fact that they include both public and private sources. Investigative reporting since 2023 has shown that large language models have been trained on a mix of licensed data, publicly available data such as books and internet content, and data obtained through agreements with publishers. For business owners, this lack of transparency creates risk: if you cannot see what went into the model, you cannot know whether the results it produces are influenced by biased material, rely on copyrighted content, or inadvertently incorporate personal information that should not have been there in the first place.

This question of "what's in the dataset" is not going away. Regulators are taking notice. The EU AI Act requires documentation of training data for high-risk systems such as hiring and financial services, and in the United

States, the Federal Trade Commission has already raised concerns about companies scraping personal data to train AI models without consent. For small and midsize businesses, this means the choice of vendor matters more than ever: you need to know not just what the system can do, but also whether you are legally safe using its outputs.

Think about what happens when you use AI tools internally. If you upload your customer lists to train a chatbot, or feed your financial data into an AI forecasting tool, you are creating a dataset of your own. Depending on the vendor's terms of service, that dataset may remain private to you or may be used to improve the vendor's model more broadly. That difference is significant.

Imagine sending your child to school with a carefully packed lunch every day. If the teacher gives that lunch only to your child, your child is nourished in the way you intended and you have control over their diet. If the teacher empties all the lunches onto one big tray so the whole class can share, your child will still be fed, but you have lost control of what they eat and, without your consent, your groceries are feeding all the other children. That is essentially the difference between private and shared datasets.

Similarly, if your proprietary data helps the vendor make its product smarter for everyone, you may have just given away some of your competitive advantage. This is why contracts and disclosures are as important as the tools themselves.

Types of Learning

There are three main ways machines learn: supervised learning, unsupervised learning, and reinforcement learning. Each method is powerful in its own way, and each comes with limitations you need to understand.

Supervised Learning

Supervised learning is a type of machine learning in which the program is trained on a labeled dataset. In other words, the system is given examples that already have the right answers attached, and it learns to generalize from them.

For example: Imagine you want to build a system that can classify images as either dog or cat. You provide a dataset packed with labeled examples: pictures of dogs marked "dog" and pictures of cats marked "cat." The system studies the data, makes predictions, and receives constant feedback about whether it was right or wrong until it gets better at distinguishing the two.

This is like teaching a child to sort laundry by giving them clear instructions: "socks in this pile, shirts in that pile." At first the child may confuse a crew sock with a knee-sock, but with repeated correction and clear labels, the child improves.

Supervised learning is especially useful when you want predictions based on historical patterns. Businesses use it for things like forecasting demand, scoring sales leads, detecting fraud, or predicting whether a customer is likely to defect. For SMBs, think about your sales history: if you have a labeled dataset of "won" and "lost" deals, you can train a model to predict which prospects are most likely to close. The limitation is that the model can only be as good as the labels you provide. If the data was biased or incomplete, the system will carry those flaws forward.

Unsupervised Learning

Unsupervised learning starts without a labeled dataset. The system is given raw data without any categories and is asked to find structure and patterns on its own.

For example: A company might load years of supply chain data from suppliers, sales records, delivery times and delays, and weather disruptions into an unsupervised learning system. The system might discover hidden relationships, such as a certain supplier consistently causing subtle delays under specific regional weather conditions.

Worth considering here is how often humans make bad decisions based on faulty assumptions. Like the time you thought the reason your dog kept whining to go out in the middle of the night was because he drank too much water, when in fact he had a bladder infection. In that case, your "cause" (water intake) was not the real driver; the problem was elsewhere. Humans create categories based on assumptions all the time, often incorrectly. Unsupervised learning is not bound by those assumptions, which allows the technology to find patterns we might not have even thought to test. That objectivity is powerful, but it can also lead to patterns that are meaningless or misleading.

For SMBs, unsupervised learning is often applied to customer segmentation. Instead of dividing customers by obvious labels like age or geography, an unsupervised model might reveal new clusters: perhaps "deal seekers who buy in bursts" or "quiet loyalists who spend steadily over years." This can be powerful for marketing strategy. The limitation is that not every pattern the system uncovers is useful. Managers must still apply judgment and test, or you risk chasing coincidences instead of insights.

Reinforcement Learning

Reinforcement learning is neither supervised nor unsupervised. Like unsupervised learning, it often starts with raw, unlabeled data. Like supervised learning, it has an objective. The system uses trial and error, guided by feedback in the form of rewards or penalties, to figure out the best way to achieve the goal.

For example, a logistics company might use reinforcement learning to optimize delivery routes. The system tries different combinations, measures which ones minimize cost and delivery time, and learns to repeat the winning strategies. Over time, it improves efficiency in ways that human planners might not have considered.

In this way, reinforcement learning is a lot like teaching a toddler to walk. You don't label every step "correct" or "incorrect." Instead, the toddler tries, falls, tries again, and is rewarded when they manage a few successful steps. Over time, they figure out the best patterns to stay upright and keep moving forward.

Reinforcement learning increasingly shows up in marketing automation. Platforms experiment with subject lines, timing, and customer touchpoints, in order to measure what drives better engagement. Over time, the system "learns" how to adjust campaigns automatically. The catch is that reinforcement learning optimizes for the goals you set. If you set the wrong target, for example, optimizing for clicks instead of conversions, the system will get very good at doing the wrong thing. This is why human oversight and clear goals matter.

○ ○ ○

Neural Networks

Much of what we humans refer to as *thinking* is actually just *perceiving*. Using our vision, sense of smell, taste, hearing, and touch, we process information from the world around us. We use memory to identify patterns, and muscle memory to function on autopilot, as anyone who has ever arrived home from work only to realize they cannot remember the drive can attest.

Machine learning can be done without neural networks. The advantage of neural networks is that they can teach a program to simulate the way a human brain works. Instead of eyes, ears, noses, taste buds, skin, and memories, a neural network is made up of tiny computational programs called neurons. Each neuron is responsible for recognizing or interpreting a specific feature in the data.

Let's go back to our database of cats and dogs. Some neurons may be responsible for recognizing just the shapes of eyes, ears, and noses. Other neurons may be responsible for recognizing fur color, while still others focus on fur patterns. The neurons then work together, much like your senses work together, to analyze pictures of cats and dogs and distinguish the differences between them.

Today's big advances in AI are largely due to the implementation of neural networks, which allow AI to recognize complex patterns and relationships in data, and perform tasks that were previously only possible for humans. How many neurons might there be in a neural network? A simple neural network, such as the app you use to scan and identify plants in your yard, may include hundreds of neurons. Advanced language models like GPT-4 are estimated to have been trained with around 1.8 trillion neurons (parameters). GPT-5, which has entered the market in 2025, is believed to be even larger and more complex, though companies continue to avoid publishing exact details.

Traditional machine learning could classify by color, texture, or shape, but it struggled to recognize complex images with high variation. Neural networks make it possible for AI to perform tasks like facial recognition and medical image analysis, both of which involve nearly unlimited variation. The fact that you can tell your high school classmates' faces apart after decades is a miracle of human processing. Neural networks make it possible for computers to simulate that feat, although still imperfectly.

Think of it like learning to recognize a song. You might hear just a few notes and know instantly that it's "Happy Birthday." A neural network is built to layer that kind of recognition, from the smallest detail to the whole picture, until it can say with confidence: this is a cat, this is a dog, this is a benign growth, this is a malignant one.

Neural networks are not abstract research anymore. They are already embedded in the tools you use. Retailers benefit when inventory software can scan and identify items with accuracy. Manufacturers use neural networks in predictive maintenance, catching problems with machinery before it breaks down. Even HR software increasingly uses neural networks to analyze resumes and match candidates to roles.

The takeaway is that neural networks are what make today's AI both so powerful and so opaque. Because they involve billions or even trillions of parameters, the path from input to output is difficult to trace. As a business owner, that means neural networks bring both extraordinary new capabilities and new risks. The system may be able to "see" things you never could, but it may not be able to explain its reasoning in ways you can defend. Which is why, as you will see later in this book, explainability has become one of the central concerns in AI governance.

Deep Learning

Deep learning is a subfield of machine learning. Both machine learning and deep learning involve teaching a machine to recognize patterns in data and to make predictions, reach conclusions, or choose a course of action based on those patterns.

The difference is that machine learning does not always require the use of neural networks, while deep learning always does. The "deep" in deep learning refers to the fact that these networks are made up of many hidden layers of neural networks.

Back to our example of cats and dogs: if you asked a computer to look at raw pixel data of pictures of cats and dogs, it would probably struggle to tell them apart. By adding multiple layers of interconnected neurons, the neural network can interpret the image in stages. One layer may identify basic shapes, lines, and edges. Another layer may combine those shapes into more complex features like eyes and whiskers. As the layers build, the network can begin to recognize entire faces, and eventually, the whole animal. The deeper the layers, the more accurate the interpretations become.

Anyone who has brought home a newborn from the hospital has experienced the human equivalent of deep learning. On the first day, every cry sounds the same (and stressful too!). Then, over the next few days and weeks, you add layers of experience. Soon you can tell the difference between a hungry cry, a wet-diaper cry, a tired cry, and a bored cry. Three weeks later, when your mother-in-law comes to visit, she hears the baby cry and confidently says, "Oh, the baby is hungry!" You know, however, that the baby is tired, because you have developed layers of knowledge that are not yet available to her. That is deep learning: adding layer upon layer of recognition until the system can make fine distinctions.

You also do the human version of deep learning when you learn to play piano or guitar. At first you only recognize the individual notes. With practice, you start to hear chords, then progressions, then eventually entire songs. Your brain builds hidden layers of knowledge until what once sounded like noise now sounds like Mozart or Paganini.

Deep learning is already behind many of the tools you use every day, even if you don't realize it. Voice assistants on your phone or smart speaker use deep learning to distinguish commands. Accounting software uses it to detect irregularities in financial records. Ecommerce platforms use it to recommend not just one product, but combinations of products that tend to be bought together. Manufacturers use it for quality control, training systems to catch subtle defects that humans might miss after hours on the production line.

Deep learning is what allows AI systems to get better with scale, but it is also what makes them harder to explain. The more layers there are, the less obvious the path from input to output becomes. For business leaders, this means deep learning is both a powerful ally and a source of risk. It can give you insights you never had before, but it can also make decisions you cannot fully unpack. And when those decisions affect your customers, your employees, or your compliance with laws, that lack of transparency can hurt you.

Natural Language Processing

Anyone who has tried to have a conversation with someone who speaks a different language knows that communication involves more than just words. Natural language processing, or NLP, is what allows computers to analyze, interpret, and generate human language in a way that is closer to the way humans communicate with one another.

Imagine an interpreter helping two people who do not share a language. One person speaks in Swahili. The interpreter listens, may ask clarifying questions to make sure they understand, and then conveys the information to the other person in Spanish. The Spanish-speaking person responds. The interpreter listens, asks clarifying questions again, and translates the communication back into Swahili.

The interpreter knows that a single word can have many different meanings, and that context changes how the word should be understood. A good interpreter does not simply match words but interprets intention, tone, and nuance. NLP works the same way, but with enormous datasets and neural networks doing the interpreting. The system analyzes written and spoken words, measures sentiment, clarifies questions, and then assembles a response that is not just a recycled sentence from its database, but a reply that is constructed to convey precisely the context and feeling required for the communication.

o o o

This is why search is changing. Type a query into Google and you still get dozens or hundreds of links, some helpful and some not. Ask a chatbot the same question and you get a single, conversational response. For many users, that feels more efficient. For Google, this is a direct threat to its entire business model.

NLP has limits. It can deliver fluent responses that sound authoritative while being completely wrong. These are the infamous hallucinations. A confident, incorrect answer can mislead customers, breaking trust faster than a pause or an "I don't know, let me find the answer for you." Many companies that rushed in 2023 to replace human customer service with chatbots learned this lesson the hard way, as customers complained, not about wait times, but about being given wrong or tone-deaf answers.

The experience at **Klarna** is a vivid example. The company rolled out an AI-powered customer service chatbot that immediately generated errors, forcing human agents to step back in and repair the damage. Even more telling was the CEO's comment that "AI can already do all of the jobs that we, as humans, do." That statement not only blurs the line between tasks and jobs, it also reveals a fundamental undervaluing of what customer service work really involves. Service agents do more than recite scripts. They listen carefully when customers are upset, recognize when a question masks a deeper problem, and notice when a product is being used in an unexpected but still reasonable way. Those acts of empathy, interpretation, and problem-solving are not just tasks; they are the essence of the job.

To position NLP effectively, business leaders need to understand that work at a deep level. AI can take care of routine responses, formatting, or triage, but it cannot attach intuitive meaning to what a customer is really saying, or adapt with the same human sensitivity when the situation is new or ambiguous. The risk for companies isn't only that chatbots will hallucinate; it's also that leaders who underestimate the complexity of roles like customer service will misuse the technology, frustrate customers, and weaken the very trust they are trying to build.

Like neural networks and deep learning, NLP is already built into your daily operations. If your CRM suggests phrasing for an email, that is NLP. If your HR system analyzes resumes for keywords, that is NLP. If your accounting software flags suspicious wording in expense reports, that is NLP. NLP can make it easier to handle routine communication, find information faster, and tailor messages to different customers. NLP also introduces risk: if you let it generate entire emails, marketing campaigns, or HR decisions without oversight, you are trusting a system that does not always know fact from fiction.

NLP is not just about language; it is about trust. You can and should use it to accelerate work, but you must not confuse fluency with accuracy. Just as you would not rely on a translator you did not trust to represent you in a high-stakes negotiation, you should not rely on NLP without human judgment layered on top.

Computer Vision

Computer vision in AI is just what it sounds like: giving eyes to a computer. It enables systems to identify objects, animals, people, actions, and even emotions on human faces.

The best way to understand computer vision is to imagine you are blind. Close your eyes and try to walk through a room. Every step is filled with risk. Add a cane (and some training) and you can navigate without bumping into things. Now imagine putting on a pair of glasses (or ear pods, or a headband, or even a bowtie; it doesn't have to literally be eyes) that can identify everything in your field of vision, tell you how far away it is, whether or not it's moving, in what direction, and even describe its color, texture, or emotional state. That is computer vision.

The magic of computer vision comes from training AI programs on vast visual datasets using neural networks. These systems learn to recognize not only shapes and patterns, but also the relationships between them. This is what allows a warehouse robot to recognize a package on a conveyor belt, or a self-driving car to distinguish between a shadow on the road and a pedestrian stepping off the curb.

Think of computer vision like a child learning to read picture books. At first, the child may only recognize bright colors and shapes. Over time, they begin to see patterns: animals, faces, actions. Eventually, they understand not just the object but the story those objects tell together. Computer vision is that process, only accelerated. Sometimes it is breathtakingly accurate, other times it's clumsy and error-prone, depending on the training and the conditions.

Computer vision is not just for tech giants. Retailers are already using it to scan shelves, identify empty spaces, and generate automatic restocking alerts. Manufacturers use it to detect microscopic defects in products, monitor assembly lines for bottlenecks, and improve safety by identifying hazards in real time. Healthcare clinics are adopting affordable imaging software that uses computer vision to flag anomalies for doctors to review. Even small restaurants are beginning to experiment with kitchen cameras that monitor food preparation for consistency and quality control.

Computer vision enhances human ability, but it should not replace it entirely. Lighting, angle, and data quality all matter. A vision system that performs flawlessly in bright warehouse conditions may fail under poor lighting. If you rely on it completely, a missed defect or misinterpreted image is not just a technical error; it can become a financial or legal liability.

There is another risk worth considering. When professionals rely heavily on computer vision, they may lose some of their own diagnostic capabilities. One recent report described medical teams who had become accustomed to using AI-driven imaging to support diagnoses; when those systems were suddenly unavailable, they realized their own skills had atrophied. This is not unique to healthcare. We already see it in everyday life with drivers who cannot navigate without GPS, or office staff who struggle when automated systems go down. To get the benefits of computer vision without the downsides, use it to support your people, not replace their ability to observe, analyze, and make decisions. These tools can help us be more effective, but over-reliance can make us vulnerable.

Computer vision is an opportunity to reduce risk and improve efficiency, but it requires the same oversight and judgment you would apply to any other part of your operations.

Robotics

Robotics involves integrating AI programs with physical systems. Robots have been used in manufacturing since the 1960s, but until recently, the assembly or production line experience of robots has not been much different from watching your Roomba bump its way around the living room furniture. Traditional robots were rigid: they had to be programmed to work with very specific and unchanging object sizes, distances of reach, and conveyor belt speeds. That meant manufacturers had to design products that reused the same packaging shapes, and they often spent vast amounts of money building dedicated production lines for each process. Change one variable, and the robots could fail spectacularly. In fact, unplanned change of a belt speed or packaging dimension could trash days or even weeks of production.

With AI and neural networks, robots can now recognize objects with greater accuracy, manipulate items of different sizes and shapes, and do so at speeds and precision levels that rival or exceed human ability. Where early robots were more like vending machines, able to repeat the same task endlessly but prone to jamming when anything unexpected occurred, today's AI-powered robots are more like short-order cooks. You can hand them different ingredients, ask for substitutions, or change the order halfway through, and they will still be able to adjust and produce a result. Not always flawlessly, but with far more flexibility than before. This is why we now see not only self-driving cars, but also warehouse robots that adapt in real time, kitchen robots that prepare food, packaging robots that adjust to different container sizes without the need to be completely reprogrammed, and surgical robots that assist doctors with delicate procedures by improving their precision and stability.

Robotics is no longer just for Fortune 500 manufacturers. Small ecommerce businesses can now lease robotic arms for packing and shipping. Restaurants are experimenting with robotic fryers, burger flippers, and even servers. Small-scale farms are testing drones that can identify weeds and pests and apply treatments only where needed. Warehouses of every size are introducing mobile robots that move goods across the floor, reducing physical strain on workers.

Robotics can bring efficiency and consistency, but it also introduces cost and complexity. Robots require training, maintenance, integration, and human oversight. For SMBs, this means the question is not just "what can the robot do," but "what will it take to keep it running, and will the return justify the investment?" In the right context, robotics can be transformative, but rushing in without a clear plan can leave you with an expensive machine sitting idle in the corner.

Explainability

By now you may be feeling both excited and uneasy about the potential of AI. This is exactly why explainability matters. AI systems operate on vast datasets, are incredibly complex, and work at superhuman speed. This makes it difficult, and sometimes impossible, to comprehend how or why they reach their conclusions. Explainability is the principle that we should be able to understand the reasoning behind the outputs of an AI system. Without explainability, we cannot evaluate whether a result is accurate, fair, or even legal.

When I first wrote this book, I asked ChatGPT a question about its own explainability. The 2023 version of the model gave an answer that was striking in its honesty. It explained that it could highlight some of the words and phrases that influenced a response, but admitted that the deeper decision-making process involved so many factors that it could not render them transparent. In other words, it could sometimes show us the breadcrumbs, but not the full trail. That hasn't changed in 2025. AI systems may produce highly persuasive answers, but they still lack the ability to tell us exactly why those answers emerged.

We understand this problem in other parts of our lives. When you hire a surgeon, you want to know where they studied and trained before trusting them with your body. When you vote, you likely take into account a candidate's past record before entrusting them with your future. In both cases, you use background and reasoning to build confidence. With AI however, if you cannot see the training data or the reasoning process, you are essentially taking advice from a black box.

For small and midsize businesses, this lack of explainability is not just a philosophical concern. It is a practical one. Imagine your HR software rejects a qualified candidate, or your credit system flags a legitimate transaction as fraud. If you cannot explain why, you cannot defend the decision to a regulator, a customer, or a court. The stakes are getting higher, because regulators are tightening the screws. The EU AI Act, passed in 2024, requires explainability for "high-risk" systems such as hiring, healthcare, and finance. In the United States, agencies like the Federal Trade Commission (FTC) and Equal Employment Opportunity Commission (EEOC) have made clear that companies will be held accountable for the outcomes of their AI systems, whether or not those outcomes are explainable.

Simply put, explainability is not optional. If you cannot trace or defend an AI system's reasoning, you should not use its outputs without human oversight. AI can be a powerful assistant, but it cannot be a trusted advisor unless its logic is transparent enough for you to take responsibility for its results.

You Now Know the Essentials

A true understanding of AI takes years of study and requires a strong foundation in computer science, mathematics, and statistics. The good news is that you do not need all of that to use AI responsibly in your business. By learning the core principles of machine learning, datasets, types of learning, neural networks, deep learning, natural language processing, computer vision, robotics, and explainability, you now have a foundation strong enough to join the conversation, question the tools being sold to you, and make informed decisions about adoption.

Think of this as learning the grammar of a language rather than becoming a poet. You may not write symphonies of code, but you can now recognize the basic structures, see when something doesn't look right, and know when you need to ask harder questions. That level of understanding is what separates a business leader who buys into hype from one who can steer a company through both the opportunities and the risks of AI.

These are more than technical risks; they are ethical and social risks as well. If the data we use to train AI is infected with bias, the outputs will be biased as well. If AI systems process personal information without consent, privacy is violated. If companies rely too heavily on AI without human oversight, people lose not just jobs but also skills. These issues are not theoretical. They are already showing up in the way companies hire, market, manage employees, and serve customers.

Which is why the next chapter looks squarely at those ethical and social concerns. Understanding the technology is only the first step. The harder work, and the work that will determine whether AI strengthens or weakens your business, is knowing how to navigate its impact on people, trust, and society.

Ethical and Social Concerns

Ethical and social debate about AI has been ongoing since the 1950s and 1960s when AI was still more science fiction than science. Early pioneers like Alan Turing and John McCarthy raised questions about the implications of machines that could think, even as they were laying the groundwork for those machines. In the 1980s the conversation intensified, as computer hardware improved enough to make some of those ideas practical. Until then, most AI concepts were purely theoretical, and when you're just discussing theory, it is easier to wave away ethical concerns.

Over the past five years the pace of development has accelerated, and with this acceleration has come urgency. The debate now centers on a set of recurring issues: bias and discrimination, privacy, accountability and transparency, job displacement, dependence on technology, data ownership and control, and the potential for harmful uses. These are not abstract problems. They are already showing up in daily business practices. Each of these concerns deserves its own book, but for our purposes we will summarize them here with an eye toward what they mean for you as a business owner.

Bias and Discrimination

The concerns about bias and discrimination in AI relate to the datasets systems are trained on, the algorithms they use, and the inherent biases of the researchers responsible for choosing the datasets, writing the algorithms, and managing the training. Data biases arise from historical data that reflects the societal values and prejudices of the time. Algorithms can be biased because of the assumptions built into them. It does not matter how unbiased a person tries to be; all humans carry inherent and hidden biases that shape the way we think and act.

The result is that AI systems can perpetuate or even amplify discrimination, making bias more deeply entrenched. We already have facial recognition systems that are less accurate for people of color, perpetuating long-standing problems with wrongful accusations. In fact, multiple city governments, including San Francisco and Boston, have limited or banned the use of facial recognition in policing because of these concerns. HR systems are increasingly using AI to screen applicants, arrange interviews, answer queries, and even suggest cultural fit within teams. If those systems are trained on biased data, the outcomes can disadvantage people of color, women, older workers, people with disabilities, names that signal ethnicity or cultural heritage, or anyone from an underrepresented background.

For small and midsize businesses, this is not a distant "big tech" problem. If your hiring software screens out qualified candidates because of bias, you are not only missing out on talent, you may also be creating legal exposure. If your customer service chatbot communicates in ways that reflect subtle stereotypes, you risk alienating customers who will no longer trust your brand. Even marketing tools that promise "personalization" can reinforce inequities by showing different pricing or product availability to different groups.

Think of it this way: if you gave a child only books from one narrow viewpoint and asked them to describe the world, they would learn fluently, but their worldview would be distorted. AI is no different. It can only reproduce the data it is trained on. If the data is biased, the results will be biased.

The impact of bias in AI is not just theoretical. It can reduce access to opportunities, exacerbate existing inequalities, and erode trust in society. The solutions are challenging but essential. They include eliminating bias where possible by removing tainted data, or solving for bias by adding data for underrepresented groups or adjusting for historical imbalances. Developers must also refine the way neurons and their connectors are programmed, improve the learning methods used in training, and make those processes transparent enough to provide true explainability.

Bias risk is addressed in a more granular way later in this book, where we review specific AI program applications. For now, the takeaway is simple but critical: AI does not remove human bias. It encodes and accelerates it unless we make deliberate efforts to prevent that outcome. For SMBs, that means asking vendors hard questions, auditing systems regularly, and ensuring there is always a human in the loop.

Privacy

AI systems are capable of collecting, analyzing, and processing vast amounts of personal data, often without individual knowledge or consent. This creates the potential for data breaches, unauthorized sharing of private information, and even surveillance. Until very recently, there was no mechanism for individuals to opt out of their personal data being included in the datasets used to train AI. That is beginning to change, though unevenly. In the European Union, the General Data Protection Regulation (GDPR) has been interpreted to give individuals more power to demand removal of their data from training sets, and the 2024 EU AI Act builds on that foundation by requiring documentation and safeguards for high-risk systems. In the United States, however, individuals still have almost no direct ability to prevent their data from being scraped or reused. California's Consumer Privacy Rights Act (CPRA) offers some limited protections, but there is no national standard yet.

For small and midsize businesses, the privacy issue is not abstract. When you use AI tools that analyze your customers' data, you are taking on responsibility for how that data is handled. If the vendor you work with suffers a breach, your customers will not draw a distinction between the tool provider and your company. It will be your brand they hold accountable.

Imagine hiring a bookkeeper who quietly made photocopies of every client record and stored them in an unlocked file cabinet at home. Even if nothing bad ever happened, the exposure alone would put your clients at risk and your business on the line. That is effectively what happens when AI vendors harvest and store data without clear safeguards.

What's a business owner to do? At a minimum, always read the vendor's privacy disclosures, demand clarity about how customer data is used, and assume that if you do not have explicit contractual assurances, your proprietary information may end up enriching the vendor's broader models. In the age of AI, privacy is one of the central risks facing businesses.

Accountability and Transparency

ChatGPT is just one AI system, but it remains the most familiar and most widely used around the world. Yet it does not disclose the datasets it was trained on, and it still cannot provide true explainability. While some AI systems, such as those built on frameworks designed specifically for transparency and interpretability, are making progress, they are still the minority. The opacity in AI development makes it nearly impossible to challenge the information these systems produce or to fully understand how decisions are reached, or whether those decisions are biased or unfair.

When systems hide their reasoning, the consequences can be severe. Individuals can be wrongly accused of crimes, denied loans, or excluded from job opportunities with no way to appeal or even understand why they were rejected. Additionally, because the reasoning is hidden inside complex networks with billions or trillions of parameters, it is difficult to hold the developers accountable for the harm their systems cause.

As of 2025, regulators are beginning to address this problem. The EU AI Act requires high-risk systems to provide documentation about how decisions are made. In the United States, the White House Executive Order on AI (2023) directed federal agencies to develop guidance for AI transparency, which set a series of agency rules in motion. Although that order was replaced in early 2025 with one more focused on innovation and competitiveness, agencies such as the FTC and EEOC have continued to make it clear that companies will be held responsible for discriminatory or harmful outcomes, regardless of whether the system itself can be explained.

For SMBs, this means you cannot simply assume that opacity is "the vendor's problem." If your business uses an AI-driven system to screen résumés, assess creditworthiness, or generate customer recommendations, you are responsible for the outputs. Imagine if a financial advisor told you they made an investment "because the computer said so." You would not accept that answer, and neither will your customers, regulators, or courts.

You can't outsource accountability. Accountable and transparent use of AI requires business owners to demand transparency from vendors, and to not use AI outputs without human review in areas where fairness, accuracy, or legal compliance matter. Accountability may rest with the developers in theory, but in practice it will fall on the businesses that deploy the tools.

All of this may sound like extra work, and in truth it is. But opting out of AI oversight today is like deciding to opt out of customer service. None of us would ever tell our customers that service is too much trouble, because we know trust depends on it. Governance is becoming the same kind of baseline requirement. The world has changed, and for small and midsize businesses, taking on this new layer of responsibility is simply part of staying competitive.

Job Displacement

AI's ability to problem-solve, produce content, and carry out routine tasks at high speed and with impressive accuracy has raised widespread concerns about job displacement. This is especially true in roles defined by repetitive or predictable work that can be automated. The economic and social consequences could include rising income inequality, reduced job security, and disproportionate impacts on marginalized or lower-skilled workers.

As business owners, it is tempting to view these changes only through the lens of efficiency ... and efficiency is attractive. Yet most western economies depend heavily on consumer spending. If automation reduces the spending power of significant groups of people, the long-term effect will not be higher efficiency, it will be a weaker customer base and economic instability.

There is also a quieter, but equally important, risk: de-skilling. Studies have shown that investment in robotics can lead companies to hire more people, but the jobs created are often lower-wage, less secure, and task-based. Instead of developing skills that lead to advancement, workers get stuck doing narrow tasks with little chance for professional growth. The result is a workforce that is technically employed, but far less capable of building careers or sustaining long-term engagement.

The squeeze is showing up first at the entry point into the workforce. Stanford researchers using ADP payroll data found that employment for younger workers in AI-exposed jobs has fallen since late 2022, with the steepest drops in roles like software development and customer support. These are the kinds of jobs that traditionally served as apprenticeships, where people learned the basics, built judgment, and earned the chance to step into higher roles.

This is the paradox we currently face. If AI takes on the repetitive, lower-value work, how do we train the people? Every profession relies on those early tasks as proving grounds. An entry-level employee doesn't just get work done; they learn by doing it. If AI removes those opportunities, leaders will have to create new ways for people to practice, make mistakes, and grow. Otherwise, you may gain efficiency now at the expense of the very talent pipeline you need for the future.

A White House report, "The Impact of Artificial Intelligence on the Future of Workforces in the European Union and the United States of America," concluded that "the primary risk of AI to the workforce is in the general disruption it is likely to cause workers, whether they find that their jobs are newly automated or that their job design has fundamentally changed." That observation has only proven more accurate as AI adoption has accelerated.

For business owners, the issue is not whether AI will displace jobs on a macroeconomic level, a debate that will continue for years, but how it will reshape the work inside your own business. If AI reduces the need for entry-level work, how will new employees gain the experience they need to advance? If AI speeds up certain tasks, will you expect staff to just take on more work, or will you create opportunities for them to build higher-value skills? Companies that think carefully about these questions will be better positioned to build loyal, skilled teams, while those that view AI only as a cost-cutting tool may find themselves with higher turnover and weaker talent over time.

Dependence on Technology

In one episode of the NBC television show *Chicago Med*, the character Crockett Marcel suddenly, and in the middle of a difficult surgery, found himself unable to trust his own decision-making abilities. He had been leaning heavily on the recommendations of an AI surgical analysis system. Stripped of that guidance, he doubted his instincts. The scene may have been written for drama, but the tension it highlights is real.

In everyday life, we already have a generation of drivers who cannot read maps, nor can they navigate without the aid of a GPS system. During the pandemic, when hospital power systems failed in several cities, medical staff discovered that without functioning computers they struggled to manage patient records manually or perform treatments without the computer-guided protocols and decision-support systems they had come to depend on. These are stark reminders that becoming overly reliant on technology can erode the very skills we once considered basic.

The same story is playing out in other professions. In 2024 a series of reports described doctors and radiologists who had become so accustomed to AI diagnostic tools that when those systems were unavailable, they found themselves less confident in their ability to read images. In finance, junior analysts have reported difficulty building models from scratch after relying on automated forecasting tools. In education, teachers who leaned on AI lesson-planning software admitted they felt less adept at creating original curricula when required to do so unaided.

For SMBs, these risks are closer than you might think. If your customer service staff rely exclusively on a chatbot to handle queries, what happens when the system goes down and they must resolve problems themselves? If your sales team depends entirely on AI-generated outreach, how sharp will their communication skills remain when a personalized, human response is required? If your bookkeeper always trusts the AI to categorize expenses, how well will they understand the financial picture when they need to step in manually?

The lesson is not to avoid AI, since it is far too useful for that. The real challenge is to use it in a way that improves, rather than dulls, human ability. Tools should complement human judgment, not replace it. A generation that cannot tell the difference between accurate and inaccurate information is already struggling with critical thinking; so as we increase our dependence on AI, we risk worsening that problem. Businesses that are mindful of this balance will not only protect their resilience in times of disruption, they will also ensure their people continue to grow as thinkers and decision-makers, not just as machine operators.

Data Ownership and Control

In addition to the privacy concerns already mentioned, and the fact that individuals and companies have no real way to assert ownership of their own data once it has been scraped into training sets, AI systems can easily reinforce existing power imbalances. Companies with access to large volumes of data have a structural advantage over those that do not, and the implications of this are far-reaching.

Data is the fuel of AI. The more of it you have, the more powerful your models become. Tech giants can afford to buy enormous datasets, negotiate licensing agreements with publishers, and mine user interactions for insights at global scale. Small businesses, independent designers, and individual creators simply cannot match that volume of information. The imbalance means those with the most data can train the best models; and those models, in turn, attract more users, generating even more data. It is a feedback loop where the rich get richer.

For independent artists and small creative businesses, this imbalance becomes particularly painful. If their work is scraped into training data without consent, it may power models that compete against them directly, producing "in their style" on demand; yet those artists rarely have the capital or legal resources to pursue claims of infringement. Even when lawsuits are filed, the process is slow, costly, and tilted toward corporations with deep pockets.

For small and midsize businesses outside the creative industries, the imbalance shows up in competitiveness. A small retailer may feed customer purchase history into an AI tool to gain insight, but the vendor providing the tool may be collecting data from thousands of similar businesses. The vendor now holds a dataset exponentially more powerful than any individual retailer controls. In effect, the platform owner, not the user, gains the real competitive advantage.

Think of it this way: if every small bakery in town shares recipes with a central supplier, the supplier eventually knows more about baking than any one bakery ever could. The supplier can use that knowledge to set terms, raise prices, or even open their own bakery with insights none of the others can match. This is how data ownership becomes a power issue. It is not just about privacy or copyright, it is about leverage.

Artists Sarah Andersen (creator of Sarah's Scribbles), Kelly McKernan, and Karla Ortiz recognized this imbalance when they filed a lawsuit in 2023 against Stability AI Ltd., Midjourney Inc., and DeviantArt Inc., claiming the companies infringed their copyrights using billions of copyrighted works to train image-generating systems. The case has moved slowly. On August 12, 2024, the court dismissed many of the original claims but allowed direct copyright infringement, and parts of trademark, trade dress, and inducement claims, to proceed. A second amended complaint was filed in October 2024 and the defendants responded in December. As of September 2025, the case is still in discovery.

Here are two possible outcomes for this case, based on how the courts draw analogies:

- Not infringing: Imagine hiring an artist to paint your portrait in the style of Andy Warhol. That person studies Warhol's work and recreates the style. This is legal, because "style" itself is not copyrighted. If the court sees AI training as analogous to this, the plaintiffs may lose.

- Infringing: Imagine an artist who photographs Warhol paintings, cuts them up, and assembles the pieces into a new image in the form of a collage. That is clearly infringing. If AI training is treated as the direct reuse of copyrighted elements, artists are more likely to win.

It's like your grandmother's secret peach pie recipe. If you share it with a friend who makes it at home, no harm done. The problem comes when that friend publishes it in a cookbook. Suddenly, everyone is baking your grand-mother's pie, you have lost something you cannot get back. That is what can happen when business data is used to improve shared models.

For small and midsize businesses, there are three practi-cal steps that matter most. First, always read the vendor's contracts and terms of service carefully. Phrases like "we may use your data to improve our services" usually mean your business information will become part of a shared pool, benefiting the vendor more than you. Second, learn to distinguish between ordinary operational data, like invoices, scheduling, basic customer interactions, and your important proprietary data. Proprietary formulas, training materials, customer insights, or competitive strategies should be treated as assets that rarely, if ever, leave your control. Third, ask what happens to your data when you stop using a tool. Does the vendor delete it? Keep it for their own models? Share it with partners? Ownership is not just about what happens while you are a customer; it is also about what happens when the relationship ends.

Unless you have clear contractual assurances about how your data will be used and when it will be deleted, assume that once you put it into a system, you no longer own it in any meaningful way. This does not mean avoiding AI tools altogether, but it does mean you must distinguish between the data you can safely share and the data that should never leave your walls. Businesses that make this distinction carefully will be better positioned to benefit from AI without sacrificing their competitive edge.

AI Potential for Harm

In addition to all the good and exciting things it can do, AI can just as easily be used for surveillance, discrimination, manipulation, extortion, and even weaponized for military purposes or cyberattacks. The difficulty of identifying who is using AI and for what purposes makes governance very difficult, especially when the tools themselves are widely available and often open source.

Over the past two years, examples of harmful use have moved from speculation to reality. In 2024, several U.S. school districts reported incidents where students and parents received extortion emails accompanied by AI-generated deepfake audio of their child's voice, claiming the child had been kidnapped. In the same year, political campaigns around the world experimented with AI-generated robocalls and deepfake videos designed to impersonate candidates, raising questions about election integrity. By early 2025, cybersecurity firms reported that AI-powered phishing attacks had become nearly indistinguishable from legitimate communications, with some attackers even tailoring scams to specific small businesses by mimicking the tone and style of their owners' emails.

For small and midsize businesses, these risks are not theoretical. Imagine a deepfake video appearing on social media that seems to show you or one of your employees making offensive remarks. Even if quickly debunked, the reputational harm can be lasting. Or consider a ransomware attack where AI is used to probe your systems for vulnerabilities at speeds no human hacker could match. Small businesses are already prime targets for cybercrime because they generally lack robust defenses, and AI makes those attacks faster, cheaper, and harder to spot.

Governments are deploying AI for surveillance at unprecedented scales. Authoritarian regimes are using computer vision systems to monitor populations, flag protest activity, and track dissidents. While that may feel far removed from your daily operations, the global nature of business means you may one day be asked to comply with laws or contracts that require your systems to operate within such frameworks.

Think of AI as a power tool: in the hands of a carpenter, it builds a house; in the hands of someone reckless, it can cause serious harm. The very speed and scale that make AI so useful also make it dangerous when applied with malicious intent.

For business owners, the lesson is to treat AI as part of your risk landscape. That means adding AI-driven fraud and misinformation to your crisis planning. If you already have a playbook for what to do in the event of a data breach, add sections for how to respond if your business is impersonated with deepfakes, or if AI-powered scams target your employees. You may not be able to stop the misuse of AI, but you can prepare your organization to respond quickly and maintain trust with your customers when it happens.

Calls for Ethical Frameworks for AI

On March 28, 2023, more than 1,000 AI experts, financiers, and researchers published an open letter calling for an immediate pause on what they referred to as "giant AI experiments" for at least six months, with explainability at the heart of the demand. The letter warned: "Recent months have seen AI labs locked in an out-of-control race to develop and deploy ever more powerful digital minds that no one, not even their creators, can understand, predict, or reliably control." The authors argued that development should resume only when there was confidence that "their effects will be positive and their risks will be manageable."

The letter drew debate about its motivations, with some observers suggesting the signatories were less interested in pausing development than in buying time to catch up. Regardless, the concerns proved valid. Since 2023, adoption of AI has accelerated, along with new risks such as deepfake election interference, AI-assisted cyberattacks, and mounting questions about intellectual property rights.

Two days later, on March 30, 2023, the United Nations Educational, Scientific and Cultural Organization (UNESCO) urged countries to adopt its global ethical framework, first published in 2021. UNESCO argued that its "Recommendation on the Ethics of Artificial Intelligence" offered the safeguards necessary to establish a normative ethics standard for AI. By 2025, more than 50 countries had begun aligning elements of their national AI policies with that recommendation.

o o o

Of course, calls for ethical frameworks did not begin in 2023. In 2010 the "Franklin Declaration" was developed at the Asilomar Conference on Signals and Systems. In 2016 the Institute of Electrical and Electronics Engineers (IEEE) published its "Global Initiative for Ethical Considerations in AI and Autonomous Systems." In 2017 the Asilomar Conference on Beneficial AI produced a set of 23 principles now known as the "Asilomar AI Principles." In 2019 the European Union published its "Ethics Guidelines for Trustworthy AI." Many other organizations, governments, and coalitions have released similar proposals in the years since.

By 2024, these proposals began to harden into enforceable rules. The European Union adopted the AI Act, the first comprehensive law of its kind, which classifies AI systems by risk and requires transparency and accountability for high-risk applications such as hiring, healthcare, and financial services. In the United States, the White House issued an Executive Order on AI in 2023, and since 2024 federal agencies have continued to publish rules and guidance on transparency, bias testing, and consumer protection, even as the Trump administration shifted the emphasis. The Group of Seven (G7) introduced a voluntary code of conduct, and UNESCO's framework gained traction as dozens of countries moved toward adoption.

Despite differences in emphasis, these frameworks share common ground. All call for transparency, fairness, privacy, security, human oversight, and accountability. Some highlight human rights, others stress technical transparency, and still others, such as the Green AI initiative, emphasize sustainability.

The work of developing ethical frameworks is no longer theoretical. Regulators are now using these principles to define obligations, and vendors are being forced to build compliance into their products. That means the frameworks already apply to you, whether or not you have ever read them. For small and midsize businesses, this ethical conversation has matured. It is no longer optional or philosophical. It has become the regulatory reality you will be expected to operate within.

Governance & Risk
Leading Responsibly in the Age of AI

Governance Matters More Than Ever

If you've been running a business for years without a formal governance function, you're not alone. Most small and midsize companies don't have compliance officers or risk committees, and for the most part haven't needed them. AI changes that by introducing new risks.

In this book, governance means the practical oversight for AI: clear rules for how tools are chosen, tested, and used; who approves and monitors them; how data is handled; and what happens when something goes wrong. Without that kind of process, companies are exposed to legal liability, reputational damage, regulatory fines, data misuse, and biased or unethical decisions that can harm employees and customers.

AI is not just another tool, like a word processor or an email program. Traditional software has always been fully dependent on you: you type into Word and it records what you write; you draft an email and the system sends it; your ERP runs reports based only on the data you entered. You always knew what was going in and could trace exactly what came out.

AI changes that. It is woven into the rest of your software and does more than just process your inputs. The word processor now suggests rewrites. The email system segments your list based on deliverability patterns you didn't set and maybe can't even see. The ERP forecasts demand using weather data, shipping trends, or trade disruptions you may not know about or understand. AI is not just responding to you. It is generating its own inputs and influencing decisions in ways that go beyond your direct control.

What happens if a manager makes a biased hiring decision, a marketing person contributes to a customer privacy issue, or a supply chain partner fails to meet compliance standards based on decisions they made using outputs from AI-driven software? Regulators, customers, investors, and employees will all want to know the same thing: Who is accountable for how AI is being used here?

That's why governance is no longer optional. Governance isn't just paperwork. Governance is about making decisions deliberately, documenting your reasoning, and creating accountability so you can move fast without tripping over your own feet. For small and midsize businesses, good governance is a leadership responsibility. It protects you from unnecessary risk, but it also builds trust, and in uncertain economic times, trust is one of the strongest growth assets you can have.

What Is Governance?

Governance is often misunderstood as a pile of legal documents. In reality, it's much more practical. At its core, governance is the way you make and enforce decisions. It answers questions like:

- Who has the authority to approve an AI tool before it's rolled out to the team?

- How do we ensure we're not using customer data in a way that could land us in legal trouble?

- What values and ethical boundaries do we want to hold ourselves accountable to?

- Who is responsible for monitoring whether AI is working as intended, and what do we do if it isn't?

For small businesses, governance doesn't mean adding a bunch of bureaucracy. It means building practical safeguards and everyday practices so your technology innovation doesn't put your business at risk.

A Practical Roadmap for SMBs

So how do you get started? The good news is, you don't have to invent your own governance program from scratch. Clear, practical outlines from trustworthy sources already exist, and getting started is simply a matter of reviewing these guidelines and taking the first step.

Start With NIST Framework

The National Institute of Standards and Technology (NIST) is a U.S. government agency that develops widely used frameworks for managing technology and risk. If you have ever heard of the NIST Cybersecurity

Download the Guides

- Download the NIST Policy at **werx.me/nist**

- Download the NIST Quick-Start Guide for Small Enterprises at **werx.me/nist-start**

- Download our NIST Explainer, which turns the NIST policy into plain-speak, at **werx.me/nist-easy**

- Download our ISO/IEC for Beginners, which explains the standard in plain English: **werx.me/ISO-easy.**

Framework, it is the same organization. In 2023, NIST released the AI Risk Management Framework (AI RMF 1.0).

This free and practical guide helps companies of all sizes use artificial intelligence responsibly. Instead of relying on legal jargon, it lays out four simple functions: Govern, Map, Measure, and Manage. These functions can be used to create guardrails around AI, reduce risk, and build trust with customers and partners. This framework is:

- Free and easy to access.
- Designed to scale. You can start with a few simple practices and add more as you grow.
- Provides four functions (Govern, Map, Measure, Manage) that help you make AI use deliberate, documented, and accountable.
- Start with one small policy (for example, "We will not use AI to generate customer commuWnications without human review") and expand from there.

Build Up to ISO/IEC Standards

The International Organization for Standardization (ISO) and the International Electrotechnical Commission (IEC) create global standards that companies around the world use to ensure quality, safety, and accountability. When business owners hear "ISO," the first reaction is often, "That sounds complicated and expensive." Frankly, it can be. Full ISO certification involves audits, documentation, and fees that make sense for large corporations but can be overkill for a small or midsize business.

Is ISO/IEC for Small Businesses?

ISO/IEC standards are the gold standard globally, but you don't need to dive straight into certification. Think of them as a reference manual you can adapt now, and as a long-term option if your business starts serving multinational customers or industries where compliance is non-negotiable.

That said, these new standards, the ISO/IEC 42001 (AI Management Systems) and ISO/IEC 23894 (AI Risk Management), are valuable roadmaps, even if you never pursue certification. You can purchase the documents themselves (usually a few hundred dollars) and use them like a detailed checklist.

Small and midsize companies can start by reviewing the standards to see what good AI oversight looks like globally, then borrow the relevant practices: such as documenting how you choose AI tools, tracking their performance, and making sure you can explain their decisions. If your business grows into international markets, or you want to meet the expectations of enterprise buyers and

NIST vs ISO/IEC: Which is Right for You?

ASPECT	NIST AI RMF 1.0	ISO/IEC (42001,23894)
Primary Scope	U.S. government–developed, voluntary AI risk framework.	Global standards for AI management and risk.
Accessibility	Free, open, plain-language playbook with tools and checklists.	Paid standards (a few hundred dollars); more technical language.
Complexity	Designed to scale, minimal compliance expertise required.	More formal, detailed, often requires specialized compliance support.
Cost	Free to access and implement internally.	Purchase required; certification can be expensive (often tens of thousands).
Market Recognition	Strong U.S. credibility; aligned with federal regulatory direction.	Internationally recognized; valued in global supply chains and regulated sectors.
Best Fit (Small Business)	Start here; it's practical and affordable.	Use as a reference only; certification is rarely needed early on.
Best Fit (Midsize Company)	Use as a foundation; integrate into existing processes.	Consider formal alignment or certification if selling internationally or to enterprises.

regulators, you can then decide whether pursuing full ISO certification is worth the investment.

- ISO/IEC 42001 (AI Management Systems) and ISO/IEC 23894 (AI Risk Management) are globally recognized.
- Buying the standards and reading them is inexpensive (a few hundred dollars), and you can use them as a reference guide without pursuing full certification.
- Certification itself is costly and often better suited for midsize companies working with multinational clients, regulators, or investors.

Weave Governance Into Existing Processes

Putting together a governance program does not have to mean creating an entirely new set of processes. You can weave much of AI governance into the work you already do. Use existing tools such as your HR policies, vendor contracts, or ERP and CRM systems to keep track of how AI is applied in your business. Make sure someone is clearly responsible for monitoring that use; you don't need a compliance department, but you do need account-ability. Then be sure to build in regular checkpoints, even just once a quarter, to ask the simple but essential questions like, *Is our AI still working the way we want? Are we managing the risks that come with it?*

Making Governance Doable

Governance doesn't need to mean binders of policies that sit on a shelf. For small and midsize businesses, it's about building practical habits and policies that help you stay legally compliant, competitive, and trustworthy. Start with what's free (NIST RMF), layer in ISO/IEC principles if you're working globally or want to demonstrate you've set a higher bar, and build governance into the systems you already use every day.

By taking this approach, you'll be doing what most companies, even big ones, are still struggling with: putting smart oversight in place for AI. That's how you move from simply experimenting with AI to actually creating business value with it.

○ ○ ○

Legal Concerns and the Regulatory Environment

Important note before we begin

I am not a lawyer, and this chapter is not legal advice. The legal environment around AI is changing quickly. What follows is an overview meant to help you understand the landscape and the issues to discuss with qualified counsel who understand your industry and jurisdictions. Laws, agency guidance, and court decisions continue to evolve.

At the time of this writing, the regulatory environment around AI is still characterized by gaps and rapid change. Some rules are now in force, others are being phased in, and many questions are being answered through litigation that is working its way through the courts.

Let's start with what is currently known.

AI-Created Content Cannot Be Copyrighted

On March 16, 2023, the U.S. Copyright Office issued formal guidance on works containing AI-generated material. The Copyright Office reiterated a longstanding principle in U.S. law: only works of human authorship qualify for copyright. Therefore, material generated autonomously by a machine does not.

That principle was reinforced by the "Zarya of the Dawn" case and ruling. The Copyright Office limited copyright registration and protection to the human-authored text and the human selection and arrangement of text and images. But it did not grant copyright for the images produced by AI image-generation tool Midjourney.

In 2025, a federal appeals court in Washington, D.C., affirmed the denial of copyright for a purely AI-generated image in Thaler v. Perlmutter, again underscoring that human authorship is a requirement under U.S. law.

Beyond Copyright: Protecting AI Work

Though U.S. works generated entirely by AI without meaningful human authorship cannot be copyrighted, that does not leave you without options. Trade secret, contract, trademark, and database rights may still provide protection. Other countries take different approaches, and some recognize limited rights where human creativity is involved, so it is important to understand the rules where you operate. In most cases, your strongest safeguards will come from contracts and other forms of intellectual property beyond copyright.

What makes this tricky is that there is no obvious point at which human input ends and machine input begins. It's more of a continuum, so the Copyright Office has struggled to define that line. Using software for mechanical tasks such as spell-check, grammar suggestions, or layout is still clearly copyrightable.

The same is true of creative judgment applied to AI material. For example, you cannot copyright raw AI output, but if you edit, curate, or arrange it into a finished work, such as selecting images, sequencing them into a narrative, or combining text into a designed piece, that human contribution can be protected. The safest approach if you want copyright protection in the U.S. is to treat AI as a support tool, document your human role, and disclose AI use accurately in any application.

Intellectual Property Infringement

There is another risk to consider even when no copyright exists in a purely AI-generated work: if an output is substantially similar to a protected original, publishing it can still infringe the original author's rights. Ongoing litigation illustrates how these questions are being tested. In Andersen v. Stability AI, Midjourney, and DeviantArt, the court (in August 2024) dismissed several claims but still allowed key claims, including direct copyright infringement and certain trademark and trade dress theories, to proceed. The case is ongoing in 2025.

In plain English, lack of copyright in something you produce using AI does not immunize you from infringement claims if that production copies another, protected work. Under U.S. copyright law, infringement is generally "strict liability," which means you can be held liable even if you didn't intend to infringe or believed you were acting legally. Innocence may reduce potential damages, but it does not erase responsibility. This is why it's so important to review anything you publish from AI for similarity to existing works, and to look for warranties or indemnities from your software vendors as appropriate. What might this look like in real life?

- Marketing content: Suppose your marketing team uses an AI tool to generate blog posts. Before publishing, compare the AI's draft against existing web content. Tools like plagiarism checkers (Grammarly, Copyscape, Turnitin) can flag suspiciously similar passages. If the AI produces a product description that looks nearly identical to something on a competitor's website, that is a red flag, and you should not publish it without rewriting.

- Visual design: Imagine an AI image generator creates a stock photo for your ad campaign. A careful review might reveal that the generated image strongly resembles a Getty Images photo, right down to the lighting and composition. Even if the AI claims the image is "original," using it could expose you to an infringement claim. Your safeguard is to check generated images through reverse-image search tools like Google Lens or TinEye before publishing.

- Music or jingles: If you use AI to produce background music for a promotional video, listen carefully for similarities to well-known songs. An AI-generated melody that echoes a recognizable tune could trigger copyright liability. Be sure to run AI-generated music through music plagiarism detection tools like MIP-PIA, Audible Magic, or AudioLock. Documenting your review process can also help demonstrate you acted responsibly.

- Contracts with vendors: When you license AI tools, look for clauses where the vendor warrants that their outputs do not infringe on intellectual property and agrees to indemnify you if they do. For example, a software contract might state: "Vendor represents that outputs generated by the system will not infringe third-party intellectual property, and Vendor will in-demnify and hold Customer harmless from claims of infringement arising from such outputs." If you skip these warranties and indemnities, or rely on a vendor who won't stand behind their outputs, your business will likely bear the cost of defense, any settlement, and any judgment, even if the problem came from the tool. Insurance (for example, media liability or IP coverage) can help with costs, but it won't prevent a lawsuit, and it will still usually require that you show reasonable review and approval processes. Without such safeguards, if the AI copies someone else's work, you may be left to face the lawsuit alone.

Practical Rules for AI Compliance

Practice these four habits to reduce your chances of landing in court:

1. Review before you publish. Run AI outputs through plagiarism and reverse-image tools.

2. Mind your contracts. Require vendor warranties and indemnities against IP infringement.

3. Check your insurance. Make sure policies cover IP claims tied to AI use (likely Tech E&O type policies at this time, though some insurers are starting to add endorsements and riders specifically related to AI).

4. Respect data rights. Adopt "most stringent state compliance (currently California)" as your minimum privacy standard in the US.

Getting these basics right won't eliminate risk, but it will protect your business from the most common, and most costly, pitfalls.

Even if you are ultimately proven right, or can demonstrate that you exercised due diligence, defending a copyright infringement claim is expensive and can damage your brand reputation in ways that may take years to repair. For small and midsize businesses, the better strategy is to avoid ever ending up in that position.

That means three things above all: 1) review AI-generated outputs carefully before publishing, 2) use contracts to secure warranties and indemnities from vendors, 3) confirm your insurance specifically covers IP claims tied to AI-generated content, including defense costs, and 4) protect your most valuable proprietary data from being swept into someone else's model. These practices will not eliminate risk, but they will keep you out of the most avoidable messes.

Plagiarism Proofing

AI tools can save time, but they can also accidentally copy protected work. Here are a few ways to keep your business out of trouble:

- Marketing content: Run AI-generated blog posts or product descriptions through a plagiarism checker like Turnitin or Copyscape before publishing. If a passage looks suspiciously similar to a competitor's website, rewrite it before you hit "post."

- Visual design: Use reverse-image search (Google Lens, TinEye) on AI-generated images. If the system has produced something that mirrors an existing stock photo, using it could land you in legal hot water.

- Music or jingles: Be cautious with AI-generated background tracks. If a melody sounds like something you recognize, it probably sounds familiar to others too. Even unintentional resemblance can create liability. Don't rely on any one person's music knowledge! Use tools like MIPPIA or Audible Magic to do the checking for you.

- Vendor contracts: Review the fine print. Look for warranties that outputs won't infringe on copyrights or trademarks, and indemnity clauses promising the vendor will defend and cover costs if infringement claims arise. If that language isn't there, ask for it.

- Documentation: Keep a record of your reviews and checks. If you are ever questioned, being able to show you had a review process in place demonstrates that you acted responsibly.

- Insurance: Check your business insurance. General liability policies often exclude intellectual property claims, so ask your broker about coverage for copyright or trademark disputes tied to AI-generated content. Specialized media liability or IP coverage can help with defense costs if you ever face a claim.

Privacy Violations

AI systems are capable of collecting, analyzing, and processing vast amounts of personal data, often without individual knowledge or consent. This creates the potential for data breaches, unauthorized sharing of personal data, and even surveillance. Until very recently, there was no meaningful way for individuals to opt out of having their data included in the training sets used for AI models.

That is starting to change, though unevenly, and mostly outside the United States. The European Union has taken the lead by making transparency and consent central to both GDPR and the AI Act. In the U.S., however, the landscape is fragmented, with no single federal privacy law and a growing patchwork of state-level rules.

Among those state laws, California's Consumer Privacy Act (CCPA) and its amendment, the California Privacy Rights Act (CPRA), remain the most rigorous in the country. These laws give California residents privacy rights that are broader than those available to residents of other states, and they are enforced by the dedicated California Privacy Protection Agency. More than a dozen other states have passed their own privacy laws since 2023, but none exceed California's overall breadth or enforcement power.

Overview of Consumer Rights Under California Privacy Law (CPRA)

Right to know: Consumers can request disclosure of what personal information a business collects, uses, shares, or sells.

Right to delete: Consumers can request that a business delete personal information it has collected, with limited exceptions.

Right to correct: Consumers can request correction of inaccurate personal information a business holds about them.

Right to opt out of sale or sharing: Consumers can direct a business to stop selling or sharing their personal information, including for targeted advertising.

Right to limit use of sensitive information: Consumers can restrict how businesses use sensitive data (for example, health, financial, or precise geolocation).

Right to data portability: Consumers can request to receive their personal information in a format that allows them to transfer it to another entity.

Right to non-discrimination: Businesses cannot deny goods or services, charge higher prices, or provide a different level of service simply because a consumer exercised their privacy rights.

For companies that operate nationally, the simplest and safest approach is to adopt privacy policies that meet California's requirements across the board. If you treat every customer as if they are covered by California law, you reduce compliance complexity and protect yourself against the confusion of state-by-state rules (your company may also benefit from strengthened trust). If a different state enacts more strict laws in the future, the same principle applies ... align your practices to the highest bar available, and you will remain in the strongest compliance position everywhere you operate.

What does "California compliance" look like? Essentially, California compliance means giving consumers clear rights: the rights to know what data you collect, to delete or correct their data, to opt out of its sale or sharing, to limit the use of sensitive information, to be able to easily download their data, and to exercise those rights without discrimination. These should not be viewed as "just" legal rules though; they are also business practices that signal respect for your customers.

If you use AI tools that process customer data, you are responsible for ensuring that data is collected, stored, and used in line with these rights. If your vendor mishandles the data, your customers will still hold you accountable. That means you must review vendor contracts carefully, ask how data is used to train models, and confirm whether it will be deleted if you terminate the relationship. Imagine a sales manager who keeps a spreadsheet of customer credit card numbers on their laptop with no password protection. Even if no breach occurs, just storing that information in such an exposed way is an unacceptable risk. Using AI without clear privacy practices exposes you to the very same kind of risk.

Privacy cannot be an afterthought. It must be embedded in your contracts, your policies, and your oversight. Small businesses that get this right will not only avoid liability, they will build stronger, more trusted relationships with their customers.

Misinformation and Defamation

If a company publishes AI-generated content that contains false or inflammatory claims about an individual or another company, the result can be reputational harm, defamation claims, or both. Risks that once seemed theoretical are now showing up in headlines regularly.

One of the terms you will hear often in this context is *synthetic media*. Synthetic media refers to content that looks or sounds real but is created by AI rather than recorded by a person, camera, or microphone. It includes AI-generated images of people who do not exist, deepfake videos that make public figures appear to say things they never said, cloned voices that mimic real individuals, and articles generated entirely by text models. Synthetic media can be harmless when used transparently for stock photos or marketing assets, but it can be deeply damaging when used to misinform, impersonate, or manipulate.

In 2024, synthetic media entered mainstream politics. A series of AI-generated robocalls mimicking President Biden's voice told voters to stay home during primaries in New Hampshire, an event that triggered federal investigations. That same year, several small businesses were targeted with fake videos that appeared to show owners making discriminatory comments. Even though those videos were quickly debunked, the reputational damage and customer loss were immediate, and restoring trust proved nearly impossible.

Think of misinformation like a wildfire: even if you put it out quickly, there's still damage, the smoke lingers, and people remember the fire long after it was put out.

For businesses, this is not just a theoretical reputational risk. An AI-generated article could insert your company's name into a false story about fraud. A deepfake audio clip could impersonate you, your staff, or your family members in a scam. Even automated reviews that are written fluently by AI but contain false claims can harm your brand before you have time to respond.

Businesses cannot eliminate the risk of misinformation and defamation, but they can prepare. This means monitoring for unauthorized use of your name, logo, or likeness; creating a clear crisis response plan for social media; and training staff to recognize and respond quickly to deepfake threats or AI-generated review fraud. Having a plan in place is the difference between being caught off guard and demonstrating to customers that you can be trusted even under pressure.

AI and Discrimination Mean Legal Risk

Even though political language around "diversity, equity, and inclusion" has softened, the laws prohibiting discrimination in hiring have not changed. Regulators and courts continue to apply those statutes to AI systems. If an algorithm produces biased outcomes, your company can face the same investigations, lawsuits, and penalties you would if a human manager had discriminated. Test for bias, and keep a human in the loop.

Discrimination

If AI tools used in hiring, lending, housing, or consumer services produce discriminatory outcomes, you may face enforcement under existing civil rights and consumer protection laws. Even though the Trump administration has pulled back from Diversity, Equity, and Inclusion (DEI) initiatives and softened the language of "equity" in federal guidance, the underlying statutes have not changed. Courts and regulators can and do apply those laws to AI systems, and state legislatures are increasingly active as well.

For businesses, the practical responsibility remains the same: test your systems for bias, keep a human in the loop for consequential decisions, and be prepared to defend the fairness of your processes. This isn't just about compliance. Diverse teams and fair treatment strengthen decision-making, innovation, and customer trust. If even the *perception* of discrimination creeps into your hiring or customer-facing systems, it can damage your reputation with customers and employees, and it can damage your business through lawsuits, regulatory action, or lost opportunities.

Regulatory Compliance

Until recently, most AI rules were voluntary, more like codes of conduct or best practice guidelines than laws. Now, with the EU AI Act in force and U.S. agencies issuing binding guidance under existing statutes, we are starting to see enforceable obligations: rules with legal consequences if you fail to comply.

United States: The U.S. does not yet have a single national AI law like the EU's AI Act. Instead, the federal government is using existing laws and agency powers, plus executive orders, to shape how AI is managed. In 2023, the White House issued an Executive Order on AI that directed agencies to create safety tests, set transparency standards, and strengthen protections against bias and fraud.

In early 2025, the Trump administration replaced that order with one that puts more emphasis on U.S. leadership and innovation, but the agencies still have to enforce existing consumer protection, civil rights, and privacy laws, and they are applying those laws to AI.

What this means for businesses is that you cannot assume AI is unregulated. If an AI tool you use produces biased hiring results, the EEOC can still investigate under employment discrimination law. If an AI-generated advertisement makes false claims, the FTC can still act under consumer protection law. And if your use of AI leads to a data breach, state privacy laws, particularly California's, will apply. Federal agencies are continuing to publish new rules and guidance, so the regulatory net is tightening, even if there is not yet one comprehensive AI law.

European Union: The EU AI Act is now law, and its rules are being rolled out in stages. Some uses of AI, such as AI that manipulates people in harmful ways, were banned altogether, and the rules entered into force in February 2025. Rules for "general-purpose AI models" (the kinds of big models behind tools like ChatGPT) started in August 2025. Rules for "high-risk systems," such as AI used in hiring, finance, healthcare, or critical infrastructure, begin in August 2026, and more requirements will be added over time. The European Commission has confirmed that these deadlines will not be delayed.

If you sell products or services in the EU that use AI, you should be ready for record-keeping, transparency about how your systems work, and ongoing monitoring to show they are safe and fair, with stricter rules depending on how risky your system is judged to be.

This book focuses on the U.S. and EU because they set the most consequential rules for global business. Other countries, including the U.K., Canada, Brazil, and China, are developing frameworks as well, but most follow or react to U.S. and EU models. If your company operates in those jurisdictions, you should track local developments in addition to what's covered here.

Other Countries to Watch

While the U.S. and EU set the pace for AI regulation, other governments are moving quickly: If you do business in these markets, monitor local rules closely.

- United Kingdom: Developing a pro-innovation AI framework that relies on existing regulators rather than a single AI law.

- Canada: The proposed Artificial Intelligence and Data Act (AIDA) will regulate high-impact AI systems, with an emphasis on transparency and accountability.

- Brazil: Considering AI legislation modeled partly on the EU AI Act, with strong consumer and data protections.

- China: Already regulates AI-generated content and algorithms more tightly than most, requiring registration and government review for some systems.

Enforceable rules are here, they are tightening, and they increasingly expect businesses to show consistent oversight across all users, devices, and platforms. The White House and U.S. agencies continue to emphasize transparency, testing, and consumer protection, even without a single national AI law. In Europe, the EU AI Act has moved from guidance to binding law with staged deadlines and significant obligations. Other countries, including the U.K., Canada, Brazil, and China, are advancing their own frameworks, many of them modeled on or responding to U.S. and EU approaches.

Business owners should not assume AI is unregulated. The existing laws on privacy, consumer protection, civil rights, cybersecurity, advertising, and intellectual property already apply to AI, and new frameworks are adding further requirements. AI is now firmly part of the legal and compliance environment, and regulators are enforcing it.

Leveling the Field

For small and midsize companies, the technical and financial barriers to adopting AI are real. Too often, tools and training are designed with large enterprises in mind, leaving smaller firms struggling to keep up. That's why small business advocacy is essential.

Advocacy ensures that the voices of small business owners are heard in policy conversations about technology, innovation, and economic growth. Without advocacy, decisions about funding, training, regulation, and access to markets are made with only the needs of large corporations in mind. With it, lawmakers are reminded that small businesses create nearly half of private-sector jobs in the U.S. and that their ability to adopt and benefit from new technologies directly impacts national competitiveness.

Initiatives like the *Goldman Sachs 10,000 Small Businesses Voices* program are working to close the gap by pushing for policies that:

- Expand digital literacy so entrepreneurs and employees can use new tools confidently.

- Improve access to affordable technology that levels the playing field between small firms and large corporations.

- Open financing and training programs that give smaller companies the resources to innovate and grow.

- Encourage fair competition so that new entrants and smaller players have a seat at the table alongside industry giants.

When small and midsize businesses engage in advocacy they help shape an environment that determines whether they can compete and thrive in the AI era.

View 10ksb Voices policy recommendations at **werx.me/voices**

Andrea Hill is a member of the Goldman Sachs 10,000 Small Businesses Voices community and serves on select committees within the initiative.

o o o

AI and Cybersecurity

Most of this book is about how your company uses AI: the systems you adopt, the policies you write, the way you manage employees and customers. This chapter is different. Here, we're (mostly) talking about how other people's AI can affect you.

Cybersecurity is the area where other people's AI can cause you the most harm. Attackers now use AI to create more convincing scams, to automate break-in attempts, and to spread misinformation about your business at lightning speed. Even if you never adopted AI at all, you'd still be exposed to these risks.

That's why small businesses can't delay building real cybersecurity practices any longer. The threat has outpaced the old "we're too small to be a target" mindset. AI has put sophisticated attack tools into the hands of anyone with an internet connection, and that changes the game for every business.

Small and midsize businesses have always been targets for cybercrime. Attackers know you don't have the security budgets of Fortune 500 companies, and they count on the fact that many SMBs delay or underinvest in protection. That was already risky before AI. Now, it's downright dangerous.

AI has also changed the *nature* of attacks. Phishing emails no longer read like clumsy scams; they mimic the tone and style of real colleagues. Voice cloning can produce convincing voicemails from the "CEO" asking someone in accounting to wire funds. Deepfake videos can impersonate staff or leaders with frightening realism. Automated bots can probe your systems for weak passwords or outdated software at speeds no human can match.

We're at the point where SMBs cannot afford to delay cybersecurity any longer. If you've been putting it off, AI should be the wake-up call.

Where to Start: Policy and People

The first step is to realize you don't have to invent a cybersecurity policy from scratch. Two globally recognized frameworks already exist:

- NIST Cybersecurity Framework. This is practical guidance from the U.S. National Institute of Standards and Technology. It breaks security down into five core functions: Identify, Protect, Detect, Respond, and Recover. You can download it free and use it as a checklist to get started.

Quick Cybersecurity Wins

- Require MFA on all systems.
- Replace software built on outdated technology.
- Update all software regularly and apply security patches as released.
- Train employees to recognize phishing and voice clones.
- Ban sensitive data from being uploaded into AI tools.
- Set up a clear reporting process for suspicious activity.
- Back up critical data regularly and test your recovery process.

- ISO/IEC 27001. This is the international standard for information security management. Full certification is expensive, but buying the document gives you a roadmap of what "good security" looks like, and you can adopt the practices that make sense for your business and size.

If you don't have an internal IT department, you still have options. Many SMBs outsource cybersecurity to managed service providers (MSPs) or IT consultants. Look for providers or individuals with recognized certifications such as:

- CISSP (Certified Information Systems Security Professional)
- CISM (Certified Information Security Manager)
- CompTIA Security+ (a good entry-level credential)

These certifications don't guarantee perfection, but they show the person has been trained in security principles and best practices.

You don't necessarily need a full-time chief security officer, but you do need someone, whether internal or external, who is accountable for protecting your systems and data. Cybersecurity can't be "everybody's job." If everybody is in charge, nobody is in charge.

The Human Factor

Technology is only as strong as the people using it. Most breaches don't start with a sophisticated hack. They usually start with someone clicking a link they shouldn't, reusing a weak password, or uploading data to an unsafe program.

AI makes these tricks much easier to fall for. Phishing emails now sound like they came from someone you know. Voice clones can sound exactly like a trusted co-worker. If your team isn't trained to pause, question, and verify, even the best systems won't protect you.

That's why employee training is the single most important part of cybersecurity. Teaching people to slow down, spot red flags, and report suspicious activity will do more for your safety than anything else you do.

Risks to Watch

AI brings new twists to familiar threats:

- AI-powered phishing. Emails, texts, or chat messages that look and sound real. They may even include personal details scraped from social media.
- Deepfakes. Audio or video impersonations that can damage reputations or trick employees into harmful actions.
- Automated attacks. Bots that scan for weak passwords, outdated software, or unprotected ports.
- Data leaks. This risk often comes from inside your own business. Employees may paste sensitive data

into AI tools without realizing that it could be stored, shared, or even used to train someone else's model.

Practical Defenses

Here are steps every company of every size should take:

- Train staff. Show them how to spot phishing attempts, voice clones, and deepfakes. Make reporting easy and encourage a "better safe than sorry" approach.

- Use MFA (multi-factor authentication). Require it on every system your company uses, including email, CRM, accounting, payroll, and cloud storage.

- Because identity has become the primary battleground for attackers, businesses need strong systems to manage who can log in, what they can see, and what they can do. This is called *identity and access management*, and it is the digital equivalent of handing out the right keys to the right people ... and keeping everyone else out.

- Control data uploads. Set clear rules for what data can and cannot be pasted into AI tools. Treat customer lists, financials, and proprietary information as off-limits.

- Stay current. Apply software patches, update firewalls, and replace unsupported systems. Most successful attacks exploit outdated, unpatched software.

- Monitor activity. Use monitoring tools to flag unusual logins, large file transfers, or suspicious traffic. Also use monitoring software to pay attention to online mentions of your company that may be fake.

Cybersecurity Provider Checklist

When evaluating an MSP or consultant, ask:

☐ What certifications do you and your team hold (CISSP, CISM, CompTIA Security+)?

☐ Do you follow NIST or ISO standards in your work?

☐ How do you monitor for threats, and how often do you review defenses?

☐ What's your response plan if we suffer a breach?

☐ How do you keep up with new AI-driven attack methods?

If they can't answer clearly, keep looking.

- Plan for a breach. Assume you'll face an AI-driven scam or deepfake. Write down what you'll do, who you'll call, and how you'll reassure customers. A plan made in calm moments is worth ten made in crisis.

AI in Cybersecurity Applications

So now let's look at AI tools you can implement to improve your cybersecurity. AI may have made attacks more sophisticated, but it's also giving small businesses access to security tools that used to be out of reach. In fact, some of the most practical, high-return uses of AI for small and midsize businesses are happening in cybersecurity. These systems work quietly in the background, spotting patterns and anomalies that even the most vigilant human teams could easily miss.

Here are a few of the most common ways AI is showing up in cybersecurity today:

- Threat detection and monitoring. AI-powered security platforms scan millions of data points in real time, flagging unusual login activity, unexpected file transfers, or network behavior that could indicate a breach. What once required a full-time security staff can now be handled by affordable SaaS tools.

- Fraud detection. Banks and payment processors use AI to identify suspicious transactions in seconds. Many of those same capabilities are now built into small-business accounting and payment systems, helping you catch fraud attempts before any money leaves your account.

- Phishing and email protection. AI filters go beyond keyword spotting, analyzing writing style, context, and even sender behavior to catch phishing emails that look and sound like the real thing. Some systems now warn employees when an email matches the tone or timing of a known attack pattern.

- Device protection. Any device that connects to your network, including laptops, phones, tablets, cameras, and even printers, can become a possible entry point for attackers. In cybersecurity, these are referred to as *endpoints*. AI-enabled antivirus and device monitoring tools watch those endpoints for signs of malware, ransomware, or unauthorized access, and can detect problems much faster than traditional rule-based software. If a device is compromised, these tools can automatically isolate it before the infection spreads.

- Logins and Account Security. Some AI tools now help decide when to add extra login steps. For example, they may only require multi-factor authentication if a login looks unusual, such as coming from a new device or location. This approach keeps security strong without making everyday access harder for employees.

- Brand protection and reputation monitoring. New AI services scan the internet and social platforms for deepfakes, impersonation attempts, or fraudulent domains that mimic your brand. They can alert you quickly so you can get ahead of misinformation or scams before customers lose trust.

AI-powered cybersecurity tools are more accessible than ever before. Many cloud platforms bundle AI-driven monitoring and protection into affordable monthly subscriptions. These tools give you a chance to stop threats before they turn into costly breaches, and they extend the capabilities of your team without requiring massive budgets.

Cybersecurity Risks in Everyday AI Tools

Most of the AI tools we'll discuss in the next chapter, whether for marketing, sales, HR, operations, or product design, come with their own security risks. Even if the software itself is safe, how your team uses it can create exposure. Here are a few to keep in mind:

- Marketing, sales, and business systems (CRM, ERP, and related tools): Remember that most of these platforms now have AI built in. That means whenever you upload customer lists, contracts, or sales data, you may also be sending that data into an AI training environment. If the vendor reuses or shares that data, you lose control of what was yours. Always be sure to read the vendor's terms, ask what happens to your data, and tie your AI use back to the governance and data ownership practices we discussed earlier.

- HR systems: Résumés, interview notes, and employee data are highly sensitive. If an AI tool isn't secure, you could face both a data breach and a compliance investigation.

- Operations and supply chain tools: AI-driven sensors, IoT devices, and logistics platforms expand your digital footprint. Each connection is another possible entry point for attackers.

- Shadow AI adoption: Employees testing free AI apps can bypass your security controls entirely, exposing customer, strategic, or financial data without you even knowing it.

Every AI tool you use is also a cybersecurity tool, whether you think of it that way or not. Don't treat security as a separate concern. Treat it as part of your evaluation, implementation, and training process.

Think about cybersecurity the same way you might think about parking your car in a public lot. You can't make it completely theft-proof, because a determined criminal will usually find a way in. What you can do is reduce your exposure. Leaving a laptop bag or even a shopping bag in plain view invites someone to smash the window. Hackers work the same way, only their "smash and grab" is automated. AI-powered bots can scan tens of thousands of businesses every second, looking for whatever door or window is left unlocked.

That's why the goal isn't perfection; it's reducing your odds of being the easy target. Train your people so they don't hand attackers the keys. Put proven security frameworks in place. Assign someone to keep watch. The companies that do these things aren't invulnerable, but they are far less likely to be chosen as a target. And in cybersecurity, that difference is often what keeps you safe.

You're Already Using AI

Even if you have never signed up for a separate "AI tool," chances are you are already using AI every day. Many of the systems small and midsize businesses rely on now have AI features built in:

QuickBooks: AI-powered categorization of transactions, cash-flow forecasting, invoice digitization, and fraud alerts.

HubSpot: AI drafting for sales emails and marketing campaigns; content idea generation; lead scoring suggestions; customer service chat enhancements.

Odoo: AI-assisted product descriptions, website copy, and marketing content; automated invoice scanning and expense recognition; predictive tools for sales and inventory.

Keap: AI-assisted email writing and subject lines; predictive lead scoring; automated workflow suggestions; campaign optimization based on performance data.

Thryv: AI-assisted campaign copy and social captions; automated review-response drafts; personalized marketing recommendations; content scheduling and optimization tools.

Salesforce: Einstein AI provides predictive lead scoring, forecast accuracy tools, natural language queries, and automated customer insights.

In other words, if you use these systems, you are already using AI — whether you noticed it or not. The question is not whether you will adopt AI, but whether you will manage its use intentionally and responsibly.

AI Software Opportunities & Precautions

If you have come to the conclusion that AI is something to simply be avoided, think again. AI is not going away, your competitors are certainly using it, and the opportunities it presents for business improvement are worth taking seriously. It has the potential to increase efficiency and productivity, enhance customer experience, strengthen customer loyalty, help managers and executives make better decisions, and transform operations.

Tech strategist Azeem Azhar, creator of the tech analysis platform Exponential View, predicted in 2021 that AI would usher in a new era of information supply and information economy. He compared it to the joy of the 1980s internet, when being online was about the ability to talk with academics and experts around the world and access information unavailable in your own university library without waiting for an interlibrary transfer. That insight has proven correct: in 2025, information supply is already reshaping competitive advantage.

So if information supply is the new edge, who benefits? Most people immediately think of companies that create content, compete by offering customer experiences, or provide education and learning. Those industries have seen huge disruption, and that is just the beginning.

E-commerce companies benefit from extensive personalization of shopping experiences, using data from past purchases, browsing history, and customer preferences to create the digital equivalent of a skilled personal shopper. In 2025, most major retail platforms have AI-driven recommendation engines built in, and even small Shopify stores now have affordable AI plugins that deliver the kind of personalization that was once reserved for giants like Amazon.

Healthcare companies are using AI to improve patient outcomes by analyzing vast quantities of medical data, spotting patterns that human analysts might miss, and delivering bite-sized, easy-to-understand, timely content that encourages patients to follow care plans and adopt healthier behaviors. Generative AI is now also being tested in diagnostic support and triage, though these uses come with significant oversight and regulatory scrutiny.

Financial institutions are using AI not just to detect fraud and money laundering more quickly, but also to deliver personalized financial advice to clients, often through AI-augmented human advisors. Many regional banks and credit unions now rely on AI for real-time fraud monitoring. These are tools that have moved from "big bank only" applications to accessible software-as-a-service (SaaS) offerings.

Transportation companies are using AI to optimize routes and schedules, reduce fuel consumption, improve safety, and communicate delays in real time. AI-powered logistics platforms are now widely available to SMBs, with predictive models that anticipate supply chain risks from weather, health crises, or regional conflicts.

In fact, any company that uses data to make decisions and to compete effectively in the market will benefit from AI-driven insights, decision support, and recommendations. Many of today's business tools already include AI, from marketing platforms like Thryv and CRMs like Keap and HubSpot to ERPs like Odoo and NetSuite and accounting tools like QuickBooks. As a result, most SMBs are using AI whether they realize it or not. The real question is not whether you will adopt AI, but how intentionally and responsibly you will put it to work.

The following survey of AI tools that are available and affordable for businesses of every size includes both an

overview of the applications and some important guidelines and precautions for their use.

Building Customer Trust

It doesn't matter how smart your AI is if your customers
stop trusting you. People don't object to businesses
using AI; they object to being misled or treated like a
transaction instead of a person.

Building trust means:

- Be transparent. If a chatbot is answering, let customers know.
- Use AI to help, not to trick. Make sure recommendations and outreach feel useful, not pushy.
- Keep the human touch. Make it easy for customers
  to reach a real person when they need one.

Your reputation is your most valuable asset. If AI makes
your business faster but erodes trust, you've traded
away the very thing customers come to you for.

Marketing and Sales

According to recent surveys, well over 80% of marketing
professionals now report using some form of AI in their
work. AI-driven marketing has moved from "early
adopter" territory into everyday practice, and most small

businesses are already using it. AI tools for marketing remain some of the most abundant and affordable of all AI technologies.

What kinds of tools are available for marketers today?

- Content planning and ideation: AI can analyze which types of content perform best for a given audience, suggest topics, propose headlines, and generate SEO recommendations. Many SMBs now use affordable plug-ins inside tools like WordPress, HubSpot, and Google Docs to do this work automatically.

- Social media and content syndication tools: Platforms like Thryv, Hootsuite, Buffer, Later, and Canva now include AI features that generate captions, hashtags, and post ideas; recommend content based on trends; optimize publishing schedules; and provide engage-ment analytics. These tools help small businesses maintain a consistent online presence without requir-ing full-time staff.

- Writing software: Systems can produce both short and long-form text, including product descriptions, marketing emails, subject lines, social media copy, summaries, and targeted messages tuned to customer behavior. AI drafts are rarely publish-ready, and you should not use them as-is. There are both brand risks (tone, accuracy, and customer trust) and legal risks (copyright, bias, or disclosure issues). Used respon-sibly, however, these tools can accelerate the content development process by providing drafts and ideas for your team to refine into finished work.

- SEO research and analysis: AI helps marketers analyze search behavior, compare it against competitor performance, and refine content strategies. Instead of combing through keyword spreadsheets, SMBs can use tools that reveal easy opportunities to improve results.

- Customer Relationship Management (CRM) systems: CRM platforms now integrate AI to help teams manage customer interactions, personalize outreach, prioritize leads, analyze data, segment customers, and automate workflows across sales, marketing, service, and analytics. Keap, HubSpot, Salesforce, and Zoho have all added AI-powered assistants that are built into their subscriptions.

- Predictive analytics: AI can comb through customer and sales data to identify patterns and predict future outcomes for ecommerce, lead generation, and pipeline prioritization. This helps SMBs focus limited resources on the opportunities most likely to convert.

- Chatbots: Once clunky, chatbots have matured into more capable virtual assistants. AI-powered chatbots can guide customers through sales journeys, poll them about preferences, answer questions, provide product education, and even close simple transactions. For SMBs, they can deliver customer support

The Chatbot Trap: Do It Right
or Don't Do It at All

Chatbots can be a marvelous business tool, but only if they're done thoughtfully. Too many companies hand chatbot setup to IT and assume the technology will solve customer service. The tech is the easy part. The hard part is designing conversation paths that actually help customers.

- **Plan first**: Collect several weeks of customer service questions. Use CRM data if you have it.

- **Map conversation 'branches'**: Document the ways conversations can go: yes/no answers, exceptions, special cases. Every branch needs to be thought through.

- **Program with care:** Load the questions, answers, and instructions into the chatbot.

- **Test relentlessly**: Run through every possible path. Find the dead ends and frustrating diversions and fix them.

A well-structured chatbot can reduce service costs and improve sales velocity. A poorly structured chatbot magnifies inefficiency and frustrates customers. Remember Bill Gates's warning: "Automation applied to an efficient operation will magnify the efficiency. Automation applied to an inefficient operation will magnify the inefficiency."

24/7 without requiring round-the-clock staff. But chatbots also come with real risks. An AI-powered assistant cannot replace the intelligence and judgment of a trained customer service representative, and poorly planned implementations can frustrate customers and damage your brand. The technology itself is the easy part; the difficult and necessary work is mapping customer journeys and customer question-and-answer-paths, then testing those maps thoroughly. Done right, a chatbot can offset service costs and improve sales velocity. Done poorly, it will irritate customers and amplify inefficiencies.

- Marketing automation: AI extends automation by handling repetitive tasks like email campaigns, drip sequences, social media scheduling, blog syndication, lead nurturing, and cycling existing customers into new campaigns. These tasks once required large marketing teams or staff with specialized technical skills. Now they are accessible to businesses of any size.

- Customer service automation: Natural language processing enables systems to answer customer questions, make personalized recommendations, and even close simple sales transactions. Small businesses can now embed this functionality in websites and apps at relatively low cost. Chatbots are one of the most common applications, but they require careful design and oversight. If the automation frustrates, confuses, or misleads, it will damage trust faster than it saves money. Approach customer-facing automation with care: plan the interactions, test them thoroughly, and make human support readily available when customers need it.

- Sales forecasting: AI can combine historical sales data with broader market trends to produce forecasts that are far more accurate than manual spreadsheets. This helps business owners plan staffing, inventory, and financial strategies with greater confidence.

- Personalization software: AI can now improve customer engagement and increase sales by tailoring nearly every aspect of the customer journey. It recommends products or services not only based on browsing or purchase history, but also on predictive models that anticipate what a customer may want next. However, personalization can quickly cross the line from helpful to invasive. When a system feels like it "knows too much," customers may feel uncomfortable rather than cared for. So use personalization to make interactions smoother and more relevant, but always be transparent and respectful about how data is used.

What makes AI in marketing so powerful is not just the automation but the personalization. The tools can tailor experiences for each prospect or customer at a scale no human team could manage. For small and midsize businesses, the question is no longer whether to use AI in marketing and sales, but how to do so responsibly, in ways that reflect your brand voice and respect your customers' trust.

The Explosion of Content Generation Software

Content creation software has become one of the hottest areas of AI development. These tools are popular because they are both affordable and intuitive to use. The reason they are affordable is that developers can easily build them on top of existing natural language processing systems instead of starting from scratch. They do this by using connectors called APIs, or application programming interfaces. APIs are the same technology that already lets different business systems, like CRMs and ERPs, talk to each other. Because content creation tools are easy to use and inexpensive, they are often the first way small businesses experience AI.

The number of content-generation tools has exploded. In 2023 there were only a handful of well-known products; today there are hundreds, each promising to create blog posts, emails, product descriptions, ad copy, and social media campaigns at the push of a button. The abundance and affordability of these tools has made content generation the AI-entry point for most SMBs. After all, marketing is visible, and it feels easier to imagine letting a system write an email or a caption than to imagine it quietly optimizing a supply chain.

Yet ease of entry is also a problem. The majority of early AI budgets have flowed into sales and marketing pilots because they are quick to launch and easy to sell internally. Writing posts, generating subject lines, or spinning up a chatbot feels tangible and impressive. These are also the projects most likely to fail. The reason is not that the models are incapable, but that the use cases are superficial. Too often, businesses chase novelty with the attitude of "we need an AI initiative" instead of aligning AI plans with a clear strategy. The outcome is a deluge of AI-written content and scripted interactions that may look exciting at first but rarely deliver measurable value.

Because of this, businesses should take some precautions before buying into content-generation tools:

- Evaluate the quality and accuracy of outputs. AI content can sound confident while being factually wrong, tone-deaf, or off-brand.

- Be alert to bias. NLP-driven tools are especially prone to subtle stereotyping, non-inclusive language, and the reinforcement of historical bias.

- Check for strategic alignment. Make sure the tool actually solves a business problem. Don't pilot something just because it's easy to launch. Projects without a clear purpose are the quickest way to waste time and money.

- Review supplier capabilities. Ask vendors if they have invested in data privacy and security. Also, confirm they can provide the level of service your team will need. Many AI startups are very lightly staffed.

- Assess financial stability. Building and running these tools reliably at scale requires significant infrastructure. If the company does not have the capital to sustain and grow, you risk being left dependent on a tool that disappears.

Despite these risks, the opportunity is real. AI content generation can play a valuable role in ideation and development, helping you:

- Generate lists of content ideas in minutes rather than hours.

- Analyze past performance and identify new directions to explore.

- Expand ideas and themes you might not have thought of yourself.

- Check copy against positioning goals to ensure relevance and keep you on point.

- Spot new keywords or topics where your business can realistically compete.

- Produce outlines or first drafts that make it faster for your team to finish the work.

- Suggest analogies, metaphors, or thematic throughlines that spark creativity.

When you use AI this way, it becomes less of a writer and more of a coach, a brainstorming partner that helps you work faster and smarter. For small and midsize businesses, the real benefit is speed and support, not replacement. The risk comes when you treat AI output as finished work. It is still prone to errors, stereotypes, and copyright concerns, and it cannot replicate your brand's authentic voice. Think of AI as an accelerator in the brainstorming and outlining stages, while your people provide the creativity, judgment, and brand voice.

AI-Generated Content as Final Work Product

AI-generated content should almost never be used as a final work product. As we covered earlier in the discussion of copyright, you cannot own the copyright to AI-generated content, and even then, you are not protected from liability if the system has copied or imitated a protected work.

This point is especially important if your contracts require you to transfer intellectual property rights to a client. For example, if you are hired to create a marketing campaign, write training materials, or design packaging, your agreement likely assumes that the client will own the copyright in those deliverables. Because AI-generated work cannot be copyrighted, you cannot fulfill that obligation if you rely on AI for the final output. To fulfill

When Can You Use Final AI Output?

There are a few narrow cases where publishing AI-generated content without major edits can be reasonable. Outside these cases, final content should always be reviewed, edited, and finished by a human.

Generally Acceptable Uses:

- Demonstrations and examples: Showing exactly what an AI system produced, as part of teaching or illustrating a point.

- Placeholders in drafts or mockups: Using AI text or images temporarily in design comps, storyboards, or prototypes.

- Experimental or artistic projects: Making it clear that the output is intended as exploration or novelty, not authoritative content.

- Internal brainstorming tools: Keeping AI-generated slides, visuals, or text inside the team as fuel for discussion, not as finished deliverables.

Avoid Using AI as Final Work In:

- Customer-facing communications: Emails, ads, product descriptions, or social posts where brand voice and trust are critical.

- Legal, contractual, or compliance materials: Anything binding or official must be directed by humans.

- Sensitive or reputation-based content: Healthcare advice, financial guidance, or anything involving personal data or vulnerable audiences.

- Visuals that resemble stock images or artist styles: Risk of copyright infringement is too high.

A good rule to follow: if it represents your business or brand to customers, or if you would be embarrassed to admit "AI wrote this," then it is not safe to publish as-is.

This callout was AI-generated using the content originally written for this book, to provide an example of AI attribution.

your contractual obligations, final deliverables must come from human authorship.

Fun Fact (But Very Important)

Because the callout on the preceding page was produced by AI—even though it was generated solely from content originally written for this book by the author—it is not eligible for copyright protection. This does not affect the copyright status of the rest of the book, which is fully protected as the author's original work.

Beyond copyright, there are brand and trust issues at stake: content that is off-tone, inaccurate, or biased can erode credibility faster than your AI writing assistant can whip up a paragraph.

Are there use cases where AI-generated content may be acceptable as finished work? There are a few narrow ones. For example, you might use AI to generate draft alt text or image captions for accessibility, with a human reviewing them for accuracy. Another example might be using AI output as a temporary placeholder in a design mockup, or generating a visual to spark brainstorming where copyright is not a concern.

○ ○ ○

If you have a legitimate reason to publish AI-generated content that does not require copyright protection, then the following safeguards should be implemented:

- Transparency: Be clear with your audience when AI-generated text or images are being used, and explain how they were created. Make it evident if they are intended as artistic or realistic representations.

- Informed consent: If AI-generated images depict people, obtain consent from those individuals. If AI-generated text is personalized using customer preferences or behavior data, that information may be considered private and therefore requires consent.

- Attribution: Clearly attribute AI-generated content to its source, especially if you are quoting a model directly. *See the call-out box on P. 124 for an example of this.*

- Privacy: Confirm that AI-generated content does not violate privacy laws or individual rights to control their own likeness or data.

- Accuracy and representation: Test content carefully to ensure it does not perpetuate stereotypes, bias, or false claims, and that it objectively represents subject matter.

- Protection of intellectual property: Validate that AI-generated content does not infringe on others' copyrighted images, text, logos, or designs.

Treat AI as an accelerator for ideas, drafts, and iterations, but keep the final product firmly in human hands. Customers engage with your business because of its authenticity and voice. No algorithm can replicate that, and handing over your brand's voice to AI without oversight is a shortcut to both legal trouble and reputational harm.

Operations

Advances across the spectrum of business operations mean that companies can now use AI to make work more efficient, cut costs, improve safety, and increase competitive advantage. Along with these advances come legal, ethical, and workforce risks that business leaders need to think through.

AI software for operations has advanced dramatically in the past six years, and many of the most practical business benefits have shown up here, in operations, rather than in marketing. In fact, Massachusetts Institute of Technology (MIT) research has found that the biggest payoffs from AI so far have come from back-office uses like finance, procurement, and supply chain, not from marketing pilots.

Some of the most common applications in operations include:

- Robotic process automation (RPA): Software that handles repetitive tasks like entering data, processing invoices, or managing claims. A McKinsey study in 2024 found that companies using RPA at scale saved about 20–30% of the time and cost in finance and HR.

- Natural language processing (NLP): Used to automate tasks like email filtering, sorting customer service tickets, or analyzing employee feedback to spot common concerns.

- Image recognition software: In retail, cameras scan shelves to spot empty or misplaced items and generate work orders. In manufacturing, vision systems now catch defects, track equipment, flag safety hazards, and spot bottlenecks on assembly lines. Gartner reported in 2025 that manufacturers using these systems cut defects nearly in half.

- Fraud detection: AI models flag unusual transaction patterns in banking, ecommerce, and insurance. Analysts estimate that these tools are already saving billions of dollars annually by catching fraud faster than rule-based systems ever did.

- Supply chain optimization: AI forecasts demand, plans inventory, and predicts shipping delays. It can even factor in variables like weather, labor strikes, or shortages of raw materials. A Boston Consulting Group (BCG) analysis in 2024 found that companies using AI this way improved forecast accuracy by about a third and cut inventory costs by around 20%.

- Predictive maintenance: Sensors on equipment track vibration, temperature, and other signals, helping AI predict when maintenance is needed before a breakdown happens. Pricewaterhouse Coopers (PwC) research suggests AI predictive maintenance can reduce downtime by as much as 30–50% and extend the life of equipment by 20–40%.

- Inventory management systems: AI now helps businesses forecast sales more accurately, track warehouse stock in real time, and avoid both shortages and overstocks.

- Accounting software: Tools like QuickBooks and Odoo have added AI features that automate categorization, speed up reconciliation, and even highlight unusual transactions or trends in financial reports.

- Cybersecurity: IT teams are relying on AI to detect unusual activity, prevent breaches, and create security settings that are strong without blocking employees from doing their jobs. By looking at patterns across millions of events, these systems can often spot threats long before humans could.

On top of this, operations can also benefit from some of the tools we already discussed in marketing and sales, such as chatbots (used internally for HR help desks), customer service automation, and predictive analytics for staffing or resource planning.

Marketing tools tend to be about getting attention. Operations tools are about keeping the business running smoothly. They may not be as exciting as the latest image generator if you're following the Marketing Director in a "why we need budget for AI software" pitch, but oper-

AI in Finance and Accounting

AI is already showing up in your books: automated expense categorization, fraud alerts, and cash-flow forecasting are standard features in tools like QuickBooks and Odoo. These are real time-savers, but they also demand oversight. Always confirm that AI-driven forecasts match reality, and make sure someone on your team can explain the numbers. Finance is too important to leave to a black box.

ations is where companies usually see the most reliable returns.

Let's take a deeper dive into AI applications for operations through the lenses of robotics and data analysis.

Robotics and Data Analysis

AI-driven robotics and data analysis are already reshaping business operations. From machines that adapt in real time, to data models that spot patterns humans would otherwise miss, these technologies are opening possibilities that once felt like science fiction. The fact that they also tend to deliver the most reliable and measurable results among AI applications makes data and robotics two of the most exciting frontiers in business today.

Robotics

Robots used to be rigid and limited: they could repeat the same action endlessly, but if you changed the size of a part or the speed of a conveyor belt, the whole system could fail. That's why traditional robotics required expensive, dedicated assembly lines and weeks of downtime to rebuild the setup for new specifications.

AI changes this. With sensors, cameras, and neural networks, robots can now recognize objects of different shapes and sizes, adjust their grip, and adapt in real time. In practice, this means:

- Warehousing and logistics: Small and midsize companies can now lease mobile robots to move products around warehouses, reducing physical strain on workers and speeding up fulfillment. This can also reduce errors if the product data in your ERP system is clean and accurate.

- Retail and food service: Robots are being tested for shelf-stocking, cleaning, and basic food prep like frying and mixing.

- Manufacturing: Robotic arms with vision systems can switch between products without needing complete reprogramming, cutting downtime and changeover costs and increasing flexibility.

The payoff is efficiency and consistency. But robotics applications also come with costs, training, and maintenance responsibilities. For SMBs, the right question is not "what can the robot do," but "what will it take to keep it running, and does the return justify the investment."

Data Analysis

If robotics is about physical tasks, data analysis is about decision-making. AI has dramatically expanded what businesses can do with data, turning raw information into actionable insights.

Some examples:

- Predictive analytics: Instead of guessing based on last year's sales, AI can combine historical sales data with outside factors like economics, weather, regional labor trends, or supply chain disruptions to forecast demand more accurately.

- Procurement and inventory: AI can identify when suppliers are likely to run late, spot patterns in raw material price fluctuations, or suggest adjustments to purchasing schedules to save costs.

- Workforce planning: AI can help predict staffing needs by analyzing historical demand, absenteeism trends, and seasonal peaks.

- Finance and risk: AI models can scan thousands of transactions or contracts quickly to flag unusual terms or identify potential fraud.

For small and midsize businesses, this kind of analysis was historically out of reach, because it required large teams of analysts or expensive consultants. Today, many of these capabilities are embedded in systems you may already use, including ERP platforms, accounting tools, and CRMs. The tools are now available, but to get value out of them, you'll need to be asking the right questions.

As MIT research has shown, technology rarely fails because the model is weak. It fails because the use case (the way it's applied to a business problem) is weak or misaligned. Automating a bad process will not improve it; it will only help you do the wrong thing faster. However, when data analysis is tied to a clear business goal, such as reducing inventory costs or improving supplier reliability, the results can be transformative.

Retail and CPG Impacts

Retail and consumer packaged goods (CPG) companies face constant uncertainty. Demand forecasting, supply chain disruptions, and shifts in consumer behavior are influenced by unpredictable factors: weather, political movements, economic swings, and the nonstop churn of the 24-hour news cycle. These variables shape consumer sentiment, which in turn drives which products move quickly and which are left unsold.

AI helps tame some of this unpredictability. By combing through enormous datasets such as sales history, search trends, weather forecasts, logistics data, even social sentiment, AI can spot patterns that humans would miss and update predictions in real time. Instead of relying on quarterly forecasts that may already be out of date, executives can see changes as they unfold.

Visualization has also improved. Instead of waiting for static reports or dashboards, managers can now request on-the-fly insights: AI turns raw numbers into storylines with charts, infographics, and narrative explanations. This makes decisions faster and clearer, and reduces the risk of relying on gut feel and guesses when conditions shift suddenly.

AI is also reshaping the retail supply chain. From rapid product prototyping to manufacturing automation, and from burger-flipping robots to warehouse pickers, stockers, and cleaners, AI-driven robotics is creating systems that act with significant autonomy. McKinsey estimates that generative and robotics-enabled AI could add up to $400–660 billion annually in productivity improvements in retail and CPG globally.

These advances, however, risk introducing significant workforce disruption. Analysts expect continued thinning of middle management, as AI systems make routine decisions that supervisors once handled. Entry-level and junior roles are also at risk, since many basic tasks like backoffice sales support and shelf-stocking are becoming automated. HR experts warn this will reduce opportunities for professional development and career growth, which are critical to keeping employees engaged and loyal. Wages in certain roles may stagnate as automation expands, and some jobs will disappear altogether.

Yet, as with every wave of technology, new roles are also emerging. Among them:

- AI Engineers to design and refine the systems that power AI-driven robotics.

- Robotics Technicians to maintain and repair machines in stores, warehouses, and factories.

- Robot Trainers to teach robots how to perform new tasks and adapt to changing environments.

- Process Design Engineers to standardize operations across large retail networks, creating consistency and efficiency.

- Logistics Managers (an established role, but starting to look very different in practice) to orchestrate fleets of robots, drones, vehicles, and sensors that move goods.

- Customer Experience Designers to ensure the retail experience still feels human, even as more of it is automated.

- Data Scientists to make sense of the torrents of data that AI-powered systems generate.

For retail executives, the challenge is twofold: capture the benefits of emerging AI technologies, while ensuring the workforce adapts. That means not only planning for the systems themselves, but also planning for skills: reskilling existing employees, recruiting for new roles, and helping teams shift into higher-value work.

The real promise of AI is not machines replacing people, but machines and people complementing each other, or *technology as a teammate*. As Harvard Business Review put it, "Competing in the age of AI is not about being technology-driven per se — it's about building organizations that use technology to bring out the best in people."

Human Resources

According to recent data from Statista, over 30% of U.S. companies have already adopted AI-based human resource technologies, another 37% report they are starting to implement them, and about 25% say they are actively developing a plan. That means the majority of organizations are at least considering how AI will reshape their people practices.

Opportunities and Risks

Given everything we've covered so far, it's clear that HR is one of the business functions most vulnerable to bias and misinformation. Long before AI, humans have brought their own conscious and unconscious biases into hiring, promotion, and pay decisions. We see it in how résumés get skimmed for familiar schools or job titles, or how managers lean toward candidates who "feel like a good fit" because they remind them of themselves.

AI doesn't erase those problems. In fact, it can make them worse. If a hiring system is trained on historical data that already reflects biased opinions and decisions, the AI can reproduce and even reinforce those patterns, embedding them into the process in ways that are harder to see and easier to defend as "objective." That is how bias becomes further institutionalized, as human blind spots become system-wide rules.

The difference is that with AI, we at least have some ways to respond. We can test systems to see if they treat people fairly, require vendors to explain how their models were trained, and strengthen data with examples that represent candidates who might otherwise be overlooked. None of this guarantees fairness, nor does it remove the need for human responsibility, but it does give business leaders practical steps for spotting and correcting bias. Used with care, AI can help reduce the impact of bias; used carelessly, it could increase bias exponentially.

Precautions for AI Use in Human Resources

The risk of algorithmic bias in AI systems is always present, so human oversight is essential for detecting, avoiding, and correcting that risk. Organizations that adopt AI-driven human resource software should put safeguards in place from the start:

- Establish clear criteria. Make sure the algorithms were developed and tested to prioritize diversity and fairness. That includes reviewing the training datasets to confirm they are representative and not reinforcing existing bias.

- Review the decision-making process. Results should be transparent and understandable. Hiring teams should be able to see why specific candidates were flagged, selected, or rejected.

- Communicate openly. Share the criteria used for selections and recommendations with all stakeholders so that the process is visible, not mysterious.

- Audit regularly. Schedule reviews of the algorithm's performance. Look for patterns that could signal bias in the data or in outcomes, and check whether results are in alignment with your diversity and inclusion goals.

Diversity Drives Competitiveness

Research consistently shows that diverse and inclusive companies don't just look good on paper. They perform better.

A McKinsey study found that companies in the top quartile for ethnic and cultural diversity were 36% more likely to outperform on profitability.

Boston Consulting Group reported that diverse leadership teams generated 19% higher revenue from innovation.

Deloitte's research shows inclusive organizations are 2x as likely to meet financial targets and 6x more likely to be innovative and agile.

Diversity isn't only about fairness or compliance. It's a competitive advantage.

- Keep humans in the loop. AI should not make final calls on its own. Recruiters and managers need to review résumés, interview candidates, and compare their recommendations to the system's output over time to track relative success.

AI-driven HR tools can be used responsibly to improve both the quality and the diversity of the workforce. They can flag biased language in job postings, shine a light on overlooked candidates, and identify gaps in representation. They can also highlight potential bias in performance reviews and provide data that supports more equitable promotion and pay practices. None of this happens automatically. Success depends on keeping human oversight in place, ensuring the system's criteria are transparent, and continuously monitoring to be sure the system is producing fair and consistent results.

Too often, AI in hiring is being used badly. Candidates report receiving rejection notices less than a minute after uploading their résumés. That kind of instant dismissal does more than turn people off; it tells them they do not matter enough for their application to even be considered. That cannot be the best or highest use of this technology.

At minimum, companies should configure their systems to slow down those automated rejections. Ideally, they should also be looking for deeper details than where someone went to school or what job titles they have held.

Even if someone truly is not a fit, there is a difference between efficiency and disregard. If AI is going to be part of the hiring process, it should be used to strengthen both fairness and humanity, not erode them.

Enhancing Hiring Quality and Diversity

Consider an HR team facing hundreds of applications for a single role. In the past, they might have skimmed résumés for certain schools or job titles, often giving an edge to candidates with a specific pedigree. An AI-driven applicant tracking system can make the same mistakes if it's poorly designed. But used carefully, that same system can shift the focus toward actual skills and experiences that correlate with success. This opens the door to stronger candidates who might otherwise have been overlooked.

When recruitment and onboarding are done well, companies don't just fill seats; they find people who are genuinely suited to the work. This is in every business's best interest. The opportunity with AI for HR departments is not just about speeding things up, but rather, to look more expansively at talent. Too often, traditional hiring processes narrow the hiring pool in ways that have nothing to do with ability. Résumés get filtered through individual biases, interviews skim the surface instead of exploring how someone thinks or works, and selection decisions lean on preferences the hiring manager may not even be aware of. By analyzing candidates against skills and competencies instead of pedigree or personal connections, AI has the potential to counter some of those distortions.

There is growing evidence to back this up. In one study of 16,000 job applicants, AI systems were more effective than human recruiters at spotting candidates with the right skills and experience for a role, regardless of where they went to school or what their name was. Another analysis of more than 17,000 applicants by Korn Ferry found that AI tools recognized leadership potential with close to 90% accuracy, compared with about 30% for human recruiters. And talent assessment and psycho-metric testing company SHL, in reviewing over a million applicants, found that AI-driven assessments predicted job performance more reliably than traditional methods across a wide range of roles.

For small and midsize businesses, this could be a game-changer. Larger companies already have deep recruiting resources and expansive networks. AI makes some of those capabilities more accessible to smaller firms that may not have a dedicated recruiting team. The real advantage isn't just efficiency. It's the chance to hire more effectively and build a more diverse workforce at the same time.

Applications of AI in HR

AI is rapidly reshaping how companies manage their people. From hiring to retention, today's HR systems increasingly embed intelligent features that streamline processes, and, when used well, can reduce bias, and provide deeper insight into the workforce.

- Applicant Tracking Systems (ATS): AI can scan and rank resumes against job requirements, helping HR teams focus their attention on the most relevant candidates rather than wading through thousands of applications manually.

- Performance Management Systems: These platforms can monitor performance data, highlight patterns, compare results to key performance indicators (KPIs), and even suggest areas for coaching or recognition.

- Learning Management Systems (LMS): AI can recommend training courses based on an employee's role, performance gaps, or career goals; propose adaptive quizzes that focus on what someone hasn't mastered yet; and layer in gamified features that keep people engaged.

- HR Analytics: By sifting through employee surveys, retention rates, and performance metrics, AI can highlight areas of risk (e.g., teams with higher turnover) and point out opportunities for improvement in engagement or workforce planning.

- Compensation and Benefits Management: AI can benchmark salaries against the market, identify pay inequities across gender or demographic groups, and flag when pay practices may be creating risk for the company.

- Onboarding Software: Intelligent onboarding tools can deliver tailored training and resources to new employees, helping them ramp up faster and feel more connected to the organization.

Product Development

Product development departments are increasingly using AI to improve design, performance, and cost-effectiveness, and to decrease time-to-market. These software tools can help companies create innovative products that are safer, more efficient, and more affordable.

One of the greatest benefits of AI-driven product development software is that it can analyze vast quantities of data to identify patterns related to usage, consumer sentiment, lifetime value, safety, and effectiveness of products. AI can also help identify and mitigate design flaws, suboptimal materials, quality risks, and component choices that would otherwise increase the final costs and delay time-to-market.

Let's start by reviewing the most popular uses of AI in product development right now:

- Generative design software uses AI algorithms to generate rapid design iterations, allowing designers to explore a wide range of options, quickly eliminate ideas that will not work, and test ideas much earlier in the design process. In 2025, these tools are being used not only in aerospace and automotive engineering, but also by midsize manufacturers using cloud-based platforms.

- Computer-aided engineering (CAE) simulates the behavior of products in various conditions, allowing engineers to develop products that are safer, longer-lasting, and higher-performance, without building as many physical prototypes.

- Materials science software uses AI to analyze data about the properties of materials and how they may interact with one another, so design teams can more quickly weed out material combinations that are ill-advised and select the most suitable materials. Researchers are already using AI to identify new materials for safer batteries and more sustainable packaging.

- Quality control software uses AI to detect defects and anomalies in prototypes and early product runs, and to catch quality problems earlier in the process. This saves time and reduces waste.

- Tools typically used for operations, such as supply chain optimization and predictive maintenance, are also used in product development. These systems help ensure materials arrive when they are needed, and prototyping equipment stays in service without costly downtime.

AI Issues in Product Development

This category of software introduces a unique set of legal, ethical, and regulatory concerns.

- Intellectual property. The data used to train AI may include patented designs, trademarks, or trade secrets. If that data informs new product concepts, companies could find themselves with AI-generated designs that infringe on existing intellectual property.

- Safety and environmental risks. Using AI in the design process does not guarantee that the final products will be safe or environmentally responsible. AI is capable of reaching conclusions that humans might not have enough experience or data to imagine. This risk is especially high when recommendations fall outside current testing standards. Anyone who has ever invented anything knows how quickly we humans fall in love with truly innovative ideas, and how energetically we defend them. When you combine "exciting and new" with "unable to fully anticipate the outcomes," you increase the risk of unintended consequences that are expensive at best, and deadly at worst.

- Bias in design. Social, racial, gender, and cultural biases exist in product development even without AI. Lack of diversity in design teams has led to products that fit men better than women, packaging that excludes people with disabilities, and messaging missteps that reinforce stereotypes. AI does not erase these risks. In fact, it can amplify them if the data it learns from is biased. The well-documented failure of facial recognition systems to accurately identify people of color is one of the clearest reminders of how damaging biased training data can be.

Mitigating AI Issues in Product Development

Companies that implement AI-driven product development software should take the following precautions:

- Ensure the training data used to develop the algorithms is diverse, representative, and free of protected intellectual property such as patents, trademarks, trade secrets, or copyrighted material.

- Require transparency in system recommendations, and confirm that product development teams understand how and why the system made its suggestions.

- Establish clear guidelines and procedures for product development teams using AI software, including:

  - Testing requirements to ensure all recommendations are not only tested against existing standards, but flagged if they create conditions outside what can currently be tested.

- Accountability measures that define who is responsible for the ethical and legal use of these systems.

- Provide training and support to employees using AI-driven design tools so they understand both the opportunities and the risks. Training should also emphasize how to raise issues and when to call for a pause in development.

Handled thoughtfully, AI in product development can help companies reach their innovation, speed-to-market, and cost-reduction goals while minimizing potential legal, ethical, and brand risks. Used carelessly, it risks compounding the very problems businesses are trying to solve.

Data Analytics

Considering that AI is ultimately all about data, it should be no surprise that one of its most exciting applications relates to analytics and decision-making. Until recently, the computing power required for advanced analytics was affordable only to large enterprises. Most other businesses got by with some mix of spreadsheets and gut instinct (mostly the latter). In the 1990s, ERP and manufacturing resource planning (MRP) systems became more common, expanding how businesses could use data. Even then, most models were built around the assumptions of the humans interpreting them, with experience and perception shaping the results.

The introduction of AI-powered tools for data analytics changes this in fundamental ways. Analysis has become more accessible, more timely, and in many cases more intelligent than anything managers had at their disposal before.

Every AI tool described so far, whether for marketing, sales, operations, or product development, both produces and consumes data. All of that information flows into business intelligence (BI) systems. BI software isn't new, but AI advances business intelligence in ways that will be transformative to everyday business.

- Advanced analytics. BI platforms now use AI for pre-dictive modeling, clustering, and anomaly detection. These capabilities move analysis beyond the limits of human perception by finding patterns that analysts might never have thought to look for or test. That can be incredibly powerful, but it also raises a deeper challenge: when the system highlights a pattern or recommendation, we may not know what assump-tions went into it. Humans tend to question things we recognize, but when an AI program reveals something outside our experience or expertise, we may accept it at face value, even if it is built on biased or incomplete data. In other words, we don't know what we don't know, which means we may not even know what questions to ask. That is why pairing advanced analytics with human skepticism and strong review processes is so important.

- Conversational queries. Natural language processing (NLP) makes it possible for non-technical employees to ask the system questions in plain language, such as, "What were our top-selling products in July?" or "Which customers are most at risk of leaving?" Instead of spending countless hours manually reviewing and categorizing responses, users get immediate, usable answers.

- Automated preparation. AI can handle the heavy lifting of data preparation, doing tasks like data cleaning (removing errors or duplicates), normalization (putting numbers into a consistent format), and feature extraction (pulling out the most useful pieces of information). This frees up analyst time to focus on interpretation.

- Personalized experiences. BI tools are starting to adapt to each user. They can remember the kinds of questions someone has asked before, notice how that person likes information to be displayed, and present results in ways that match the user's level of comfort with data. For example, a manager who isn't trained in analytics might see a simple chart and summary sentence, while an analyst might get access to the underlying data and more complex options. When systems feel intuitive and approachable, employees are far more likely to use them for decision-making.

- Real-time insights. Instead of waiting for end-of-month reports, AI-enhanced BI tools generate alerts and updates in real time, helping managers respond quickly to shifts in customer behavior or market conditions.

- Data visualization. AI helps create visualizations that are not only easier to generate, but more interactive and easier to understand. Instead of static dashboards that take weeks to design, teams can now spin up on-the-fly charts and graphics that tell the story of the data more clearly.

One of the quiet strengths of AI in analytics is how it can compensate for the general lack of numeracy in business. Most managers and owners are intelligent, experienced people, but not everyone is trained to read charts critically or to build graphs that tell a meaningful story. Too often, data ends up as a jumble of numbers, or worse, as pretty visuals that don't add value or insight.

AI-driven analytics can close that gap by creating graphs and insights that are easier to understand and harder to misinterpret. In practice, this means people who may not be "numbers people" can still use meaningful data to make better decisions.

o o o

Precautions for Using AI in Data Analytics

It is no accident that data analytics comes at the end of this discussion. The same concerns raised in other functions, including intellectual property, transparency, accountability, bias, and discrimination, apply here as well. The difference is that with analytics, these risks can be harder to detect and manage, because analytics is a step removed from the day-to-day transactions. I call this the *data proximity problem*.

When people are issuing purchase orders, sending quotes, or recording sales in an ERP, MRP, CRM, or other system, they usually know what they are doing and can catch when something looks wrong. By the time those thousands of transactions are rolled up and analyzed in a BI tool, that clarity is gone. Flaws in the underlying data or models are harder to see, and the decisions made from those insights will carry those flaws forward. The proximity problem has always been a concern in business analytics, but AI amplifies it.

If AI-driven analytics software produces flawed insights because of biased training data, opaque logic, or the inclusion of protected intellectual property, and a manager then relies on those insights to make decisions, it can be extremely difficult to trace the outcome back to its source. This raises an unavoidable question: who is accountable when a decision leads to harm or liability?

In business, accountability matters because decisions affect customers, employees, and compliance obligations, and someone must be able to explain and defend them. The ability to trace decisions back to the assumptions behind them is essential, yet AI-powered analytics can create a black hole where those assumptions should be visible.

Of course, this isn't a brand-new problem. Humans also fail to reflect on the assumptions driving our decisions. The difference is that AI accelerates the process. Decisions can now be made faster, on a larger scale, and with more confidence, even when the foundation is flawed. That makes the potential harm bigger, faster, and harder to detect before it causes damage. Once flawed insights have been acted on, the consequences are also harder to correct, explain, or contain.

For companies that have operated without a discipline of documenting or auditing management decisions, the introduction of AI needs to be the wake-up call. Before adding analytics tools that amplify decision-making power, businesses should have processes in place that require documentation of assumptions, a chain of reasoning, and accountability for outcomes.

Data Privacy: The Elephant in the Room

The superpower of AI analytics tools is the amount of data they can process and the speed at which they work. Today's BI systems draw on massive datasets that include images, text, audio, and sensor readings. In many cases,

The Data Proximity Problem

The farther you get from the original transactions, the harder it is to spot errors. When people are working directly in systems like ERP or CRM — issuing purchase orders, sending quotes, recording sales — they usually know what "right" looks like and can catch mistakes quickly. Once those thousands of entries are rolled up and analyzed in a BI tool, the context is gone. Biases, gaps, or errors in the data are much harder to see, yet the insights built on them still drive decisions. AI makes this problem bigger, because it can generate convincing results even when the foundations are flawed.

Think of it like reading only the summary of a long conversation. The summary may be accurate, but you've lost the tone, the pauses, and the context that give it meaning. In analytics, that missing proximity is what makes it so easy for flawed data to slip through unnoticed.

that means working with hundreds of thousands or even millions of data points.

That power also creates legal and ethical risks. Datasets may include personally identifiable information (names, addresses, IDs) or sensitive material such as health records. This triggers obligations under laws such as California's CCPA/CPRA and Europe's GDPR, and many other states are beginning to adopt their own privacy frameworks.

Here are the biggest risks to prepare for:

- Data protection laws. Personal data in analytics systems can create legal liability if not handled properly.

- Consent and transparency. Individuals may not be able to give meaningful consent if data use is buried in complex algorithms and automated processes.

- Security risks. Sharing personal data with third-party analytics providers increases the risk of breaches.

- Discrimination. If training data reflects bias, analytics can reinforce inequities.

To mitigate these risks, companies should:

- Put data protection policies in place that meet the highest standards among the jurisdictions where they operate.

- Establish informed consent processes and clear transparency practices.

- Build robust cybersecurity programs, including risk assessments of vendors and strong data-protection agreements.

- Use data minimization techniques such as anonymization and pseudonymization wherever possible.

- Apply encryption, access controls, and regular audits to protect data.

- Review analytics outputs regularly to check for bias.

AI-driven analytics has the potential to transform decision-making for businesses of every size. For small and midsize businesses, the ability to ask questions in plain language and get immediate answers from real-time data is powerful. This power comes with responsibility. Decisions made on shaky assumptions, hidden biases, or poorly protected data can do more damage, more quickly, than ever before. If you're going to put AI analytics to work, put the right guardrails in place first.

Summary of AI Software Benefits and Precautions

AI technology offers tremendous potential for businesses: to streamline operations, to provide more personal and engaging customer experiences, to design and deliver innovative products, and to stand out from competitors. We have looked at how HR, operations, product development, marketing, sales, and data analytics can all use

AI-driven software in ways that make work faster, more consistent, and in many cases, more insightful. Much of this software is no longer the exclusive domain of global enterprises; it is already accessible to small and midsize companies.

These benefits come with significant responsibility. Some of the risks are already well understood, like bias, privacy, and intellectual property misuse. Others will only become clearer as the technology matures and as laws and regulations catch up. Businesses that want to use AI effectively cannot treat these risks as afterthoughts.

This means four things in practice:

1. Understand your risk profile. Know where AI touches your business and what risks are most relevant to you.

2. Put policies and procedures in place. Mitigate legal, ethical, and brand risks before they become liabilities.

3. Educate employees. Make sure staff at every level understand both the opportunities and the responsibilities of using AI tools.

4. Institutionalize responsible use. Build responsible application of AI into your culture and processes, so you are adapting as the technology evolves rather than scrambling to react.

Handled thoughtfully, AI can give small and midsize businesses advantages that were once reserved for much larger organizations. Handled carelessly, it can expose companies to risks that undermine both trust and competitiveness.

In the next section, we will outline a framework for implementing AI software successfully, adapting familiar best practices for software adoption, and adding the new procedures that AI-driven tools require.

A Cautionary Note About AI Implementation

Recent research shows that 95% of AI pilots fail. Not because the models are broken, but because the initiatives are superficial. Too many projects chase novelty instead of solving a real business problem. Too many leaders are treating AI like a "must-do experiment." They launch a few content pilots, then wonder why nothing sticks. The result is wasted resources, frustrated employees, and damaged trust in the technology itself.

AI pilots fail when they are divorced from strategy. Successful projects start with a clear business case, tie directly to measurable goals, and have leadership commitment to change management. If you're only experimenting for the sake of showing you're "doing AI," you're setting yourself up to be part of that 95%.

Your decision to implement AI-enabled technology should not be about saying "yes" or "no" to AI. It should be about making sure every "yes" is grounded in strategy, managed with intention, and designed for measurable impact.

o o o

A Framework for AI Implementation and Management

The Ripple Effect of AI Decisions

One of the things business leaders tend to underestimate is how far the impact of an AI decision spreads. AI adoption doesn't happen in a vacuum; it creates ripple effects that touch your business model, workforce design, processes, quality, customer relationships, and your competitiveness in the market.

That's why you can't treat AI like a plug-and-play feature. Choosing a tool is only the first step. You also have to think through what aspects of your business will change when that tool affects how work gets done, how customers experience your business, or how your data gets used. The technology itself may be straightforward, but the changes it triggers are proving to be challenging.

From Ripples to Reality

When every third email you open contains an announcement about a new AI software, it's tempting to click a link, launch a free trial, and start testing. That can be a fun way to spend your free time, but when it comes to making business decisions about software, it's best to follow a thoughtful selection and implementation process. I can hear your next question now ... "Even if the software is free?" Perhaps *especially* when the software is free. As the old saying goes, if you're not paying for the product, you are the product. With AI tools, that usually means your data is what's being sold, shared, or used to train someone else's model.

To protect your business from the risks of hasty software implementation, the following framework adapts familiar best practices for evaluating and managing software. It also incorporates new considerations that have emerged with the rapid rise of AI. The goal is to help you avoid risks while still making the most of the benefits AI has to offer.

What are the risks of rushing adoption?

- Learning new programs takes time. If the program doesn't ultimately meet your needs and gets abandoned, all of the hours your team invested in training and practice are wasted. Worse, that wasted time also comes at the cost of opportunity: time that could have been spent improving processes, getting your message out, or serving customers.

If You're Not Paying, You're the Product

You've probably heard the old saying: "If the product is free, you are the product." It's a reminder that free software and services usually make money in other ways, most often by collecting and monetizing your data. The idea isn't new. In 1973, artists Richard Serra and Carlota Fay Schoolman released a short film called "Television Delivers People," which argued that commercial TV's real product wasn't the shows, but the audience being sold to advertisers. The same principle applies today: when an AI tool is free, it's worth asking how your data might be used.

- Complexity as a Risk Factor. Regulators increasingly view disconnected, patchwork systems as a governance risk, since blind spots and integration gaps make it harder to prove accountability.

- Software chosen too quickly often fails to live up to expectations. When that happens, companies tend to jump to the next shiny option. Switching back and forth between programs not only creates inconsistency in the way people work, it also introduces mistakes and reduces productivity as employees are forced to constantly relearn processes.

- When teams or departments adopt different tools without coordination, they create data silos that block information from flowing across the business. Data ends up fragmented across multiple systems that don't talk to each other, and communication slows down as people spend extra time reconciling numbers rather than acting on them. The consequences down the road can be serious: lost sales, frustrated customers who feel like nobody knows their history, wasted effort, and teams accidentally working against each other instead of pulling in the same direction. An equally destructive, but less obvious, result is the small gaps that appear in the data. Missing details and slight contradictions may seem minor, but faulty data can steer people toward decisions that are misguided or, in some cases, completely wrong.

- Early-stage software frequently has gaps in its security. These vulnerabilities can expose sensitive company data and compromise your customers' data. In a business environment where customer trust is currency, one breach can undo years of careful brand building.

- Finally, many AI startups are poorly funded, which carries significant business risk. They may be acquired and folded into another company, pivot to a new direction that doesn't support your needs, or simply shut down. In those cases, you could be left with unsupported software, no future updates, and the hard choice of starting all over again with a different tool.

Any rush to software adoption puts your business at risk. Taking the time to understand your needs clearly, choose software carefully, and implement it thoroughly will ensure the software you choose meets your needs today and for years to come.

AI Complexity and Shadow Adoption

Back in March 2023, more than a dozen major corporations rolled out AI features inside their existing software platforms. By 2025, nearly every system businesses depend on, from accounting and HR to customer management, already includes AI or soon will. You may not set out to buy "AI software," but if you buy software, you are almost certainly buying AI.

This embeddedness of AI changes the way you evaluate every system. It's no longer just about whether the tool does the job you need; you also must ask what the AI features are doing behind the scenes, what data they use, and whether they introduce new legal, regulatory, or ethical risks.

AI has moved from being an optional add-on to being part of the foundation of your software environment, so every software decision now must be made with that in mind. The problem is that many businesses still run on a patchwork of disconnected software programs. When systems don't talk to each other, it's almost impossible to see where AI is active, what data it's using, or whether it's creating risk. Regulators are starting to expect companies to either unify their systems or connect them well enough to show they are in control.

Things get more complex when you add on the issue of *shadow adoption*. Shadow adoption is what happens when employees sign up for AI tools on their own because they

Why User Input Matters

When you plan for new software, every perspective counts. Executives know why the technology is needed for the business strategy. IT understands how to support it and connect it to the rest of your systems. The people who use the software every day know the details of how work really gets done, including the steps that keep customers happy and orders moving.

If their voices are left out, you risk building a system that looks good on paper but fails in practice. Remember, technology change is also cultural change. If people do not see the value or feel ownership in the process, they are far less likely to adopt the new system, no matter how powerful it is.

are trying to make their jobs easier, often with the best of intentions. The problem is that those tools sit outside your approved systems, which means data gets scattered, security may be compromised, and work can end up being duplicated or even lost.

Imagine if every person in your company decided to keep their own private version of the customer list, scattered across Google Sheets, Excel workbooks, and free trials of multiple CRMs. Eventually, nobody would know which version of the customer list was accurate. Of equal concern, tools adopted in the shadows rarely go through the legal or compliance checks you would normally require. That can leave you exposed to privacy violations, intellectual property misuse, or other regulatory problems without you even knowing it.

Software Selection

Business Requirements

The first step in any software implementation is to define the business requirements. This means identifying your specific needs and goals, then translating them into a clear, measurable statement of what you expect the software to do.

Gather input from stakeholders, including direct users, managers, interdependent teams, and executives. Then put that input together in a requirements document that spells out what the software needs to do, how it should perform, and what outcomes you expect. That document becomes your yardstick for comparing vendors and measuring success once the system is in place.

It's important to understand why each group's input matters. Executives can explain the strategic reasons a new technology is needed. Your IT team can evaluate whether the software is well written, whether it can be supported, and how it will fit with the rest of your technology stack. But it is the people who use the system every day who understand the work at the most detailed level. They know what it takes to keep customers happy, process orders accurately, and get products and services delivered. If their needs are overlooked, the system may check all the boxes on paper but still fail in practice.

Technology change is also cultural change. One of the most common reasons software projects fail is that the people who are supposed to use the new system never truly adopt it. Sometimes that happens because the software does not actually make their work easier or better. Other times it happens because they were left out of the planning and do not see why it is worth changing their habits. Including users in the requirements process not only helps you design a system that works, it also builds buy-in and increases the chances that your team will embrace the change instead of resisting it.

The Business Requirements Document should include the following:

- Functional requirements: What the software should do.
- Non-functional requirements: Qualities such as performance speed, scalability, usability.
- Technical requirements: Compatibility with existing systems, required integrations with supply chain partners, programming languages, or operating platforms.
- Business requirements: How the software will support your actual business goals.

Examples of Traditional Business Requirements

- "Improve the collection and analysis of customer feedback."
- "Reduce customer service response time by 50%."
- "Increase supply chain efficiency by automating inventory management."
- "Reduce transportation time by 15%."

Examples of AI Business Requirements

With AI-driven tools, the requirements exercise must now expand to include AI-specific expectations. These

Smart Integration: What to Look For

Not all integrations are created equal. When evaluating software, look for:

Readily achievable connections. Integrations should be supported through open APIs or standard connectors. If a system requires costly custom development or only the vendor can do the work, it will add hidden costs and limit flexibility.

Strategic connections. Integration should link the fields and functions that matter most, not everything. Connecting too much data can create clutter and confusion. Connect only the points that actually improve your processes and decision-making.

The goal is practical integration: seamless data flow where it helps, without unnecessary complexity or expense.

should reflect existing laws, anticipated regulations, and your own brand values. For example:

- "The AI features of this software were not trained on datasets with bias that could create discriminatory outcomes for employees or customers."

- "The system provides explainability that allows users to understand how and why it makes recommendations."

- "The software and what it produces are not based on protected intellectual property, such as copyrighted content, patents, or trade secrets."

- "The vendor can show legal permission to use any third-party intellectual property included in their training data."

- "The system complies with applicable data protection laws (e.g., GDPR, CPRA), including informed consent for use of personal data."

Remember that due diligence is no guarantee of protection. Under U.S. law, for example, copyright infringement is a *strict liability claim*, meaning a business can be sued for infringement even if the business had no idea its AI vendor trained the software on copyrighted material.

Asking vendors for written assurances and disclosures will not completely eliminate your legal exposure. However, widespread customer pressure should help establish expectations for transparency and accountability in the AI market, which in turn should reduce the risks of AI software presenting output that results in copyright infringement claims.

Other considerations for AI software selection:

Download a free "Business Requirements Document" template and examples at: **werx.me/brd**

- Longevity of the vendor. Many AI tools are new and unproven. Look for evidence that the company has the funding, partnerships, or financial support from a parent organization to be around in two years.

- Integration with what you already use. Any new software should be able to connect with your existing systems, such as your CRM, accounting program, or inventory software. This keeps information flowing across the business without clutter, double entry, or duplicate records.

- Model updates. AI tools evolve constantly. Ask vendors how often their models are updated, how those updates are communicated, and if you can opt out of changes that affect compliance.

- Auditability. Choose software that allows you to review and test its outputs. If a system cannot be audited, you are relying on results you cannot verify or defend, which creates unnecessary business and legal risk.

- Risk tolerance. Each business must decide how much risk it is willing to accept with AI adoption. Understanding your risk appetite up front helps you set boundaries with vendors.

Vendor Evaluation and Selection

The next stage of software evaluation is to identify companies that produce programs that could meet the expectations and goals you defined in your business requirements. Software vendors are more than just suppliers, they are long-term partners. Thoroughness in this step will lead to stronger, longer-lasting partnerships.

Vendor Due Diligence: The Essentials

Things to investigate include:

- Experience and reputation. Look for reviews, references, and case studies. Ask the vendor to connect you with current customers, and don't hesitate to follow up with those references directly.

- Financial stability. Choose a vendor that is likely to be around for the long term. Ask for financial references or evidence of funding if you are unsure. This is a reasonable and fairly common request, so don't hesitate to ask.

- Implementation services. For significant implementations, evaluate the vendor's approach to project management. How do they assess your needs? Do they have experienced staff who can work with you through planning, data migration, customization, and testing? Do they work with implementation partners or value added resellers (VARs) who are fully certified in their platform? Look for evidence of successful implementations in companies that resemble yours in size and complexity.

- Support services. Ask how support is structured and who you will be talking to first when you need help. Will you always speak with an entry-level help desk, and how can you connect with an experienced support engineer when you need them? Make sure the vendor offers support channels that fit your needs

Vendor Stability Risk

AI tools can be powerful, but what happens if your vendor gets acquired, changes strategy (and therefore, its software), or shuts down? Ask these questions up front to hedge against risk:

- Can we export our data in usable formats?
- What happens to our data if we leave?
- Are integrations built on open standards, or are we locked into one ecosystem?
- Is there a contract clause that guarantees notice if terms or pricing change?

The more vendors you depend on, the more chances there are for one of them to cause trouble. If a provider shuts down, changes its terms, or suffers a breach, you could be left exposed even if everything else is working fine. Choosing an AI vendor isn't just about features; it's about protecting your flexibility to change course without losing your data, your processes, or your customers.

(phone, email, ticket system, chat), and talk to references about responsiveness and problem-solving.

- Integration with existing systems. Determine whether the software can connect with the applications you already use, and how easy or complicated those integrations will be to create and maintain. The more fragmented your systems, the more important this becomes.

- Training. Ask about the vendor's philosophy of user adoption. Do they provide only videos and PDFs, or do they offer live sessions, online courses, or even in-person workshops if needed? Do their implementation partners provide quality training? Training is often the hidden cost of success, and vendors that skimp on training (or making quality training resources available) are setting you up for frustration and low adoption.

- Cost of ownership. Ask the vendor to be transparent about all costs, not just licensing. That includes implementation, training, annual fees, customization, maintenance, and upgrade costs. Unexpected costs can turn an affordable program into an expensive one.

Additional AI Considerations

When AI is part of the software, additional questions should be asked:

- Disclosures. Will the vendor be transparent about where training data came from, how it was obtained, and whether it included copyrighted or sensitive material? If they hesitate to answer, that's a red flag.

- Explainability support. Can the vendor provide plain language explanations, documentation, or tools that help you understand why the AI produced a particular result? Without this, you cannot defend its use to customers, regulators, or in court.

- Governance practices. Does the vendor have internal processes for ethics, transparency, and compliance? Ask how they monitor bias, test for fairness, and handle model updates.

- Privacy commitments. How does the vendor handle consent, data storage, and protection of personal or proprietary information? Get clear on whether your data will be used only for your system or fed back into their broader model.

- Philosophy and values. Does the vendor articulate a position on responsible AI development? A vendor that can't discuss its approach to ethics, fairness, and transparency is signaling that those issues are not priorities.

About Implementation Partners

If the software is sold or supported through an implementation partner (a third party that helps set up and customize the system) or a value-added reseller (VAR, a company that sells the software on behalf of software companies and provides extra services like training or support), that can be a good thing. Many software companies focus on building the core technology, then rely on partners to handle implementation, customization, and training. That division of labor can work in your favor, because a strong partner network often means you will get better service and more specialized support.

The caution is that not all partner networks are created equal. Some software companies certify and train their partners carefully, while others will let almost anyone call themselves an implementer. Before you commit, find out how the vendor manages its partner program, whether there are clear tiers or certifications, and what standards those partners have to meet. If a partner plans to bolt on third-party extensions or do custom development for you, your AI-related due diligence should extend to those add-ons as well.

It takes time to check out vendors properly, but skipping that step always costs companies more in the long run. Choosing software without vendor evaluation is like hiring an employee without checking their references, qualifications, or background. You might get lucky, but more likely you will end up with regrets. When AI is involved, those consequences can include not just wasted money, but compliance violations, brand damage, and legal liability.

Five AI Questions to Ask Before You Hit the On Switch

Before you flip the switch on a new AI-powered system, take the time to ask (and answer) these questions:

- Does it really meet our business needs? Make sure you are solving a real problem, not just chasing a shiny new tool.

- What data is it learning from? Ask the vendor where the training data came from, and whether it included copyrighted or private information.

- Can we explain its decisions? If you can't understand how the system produced a recommendation, you won't be able to defend it to a customer or regulator.

- What happens to our data? Get clarity on whether your data will stay private, be used to train the vendor's models, or be shared with others.

- What support do we really get? Find out how the vendor (or their partner) will train your team, handle problems, and keep the software updated over time.

If you can answer these questions with confidence, you are in a stronger position to launch successfully and avoid the issues that derail so many AI projects.

o o o

Implementation

Software implementation planning involves forming project teams, allocating resources to the project, creating project plans, and establishing milestones and deliverables.

Once the plan is complete and approved, the preparation phase begins. This includes data migration planning, analyzing and improving business processes, customization (keep those limited, folks), configuration, and testing (and then more testing).

AI Adds Complexity to Implementations

The introduction of AI leads to new responsibilities during planning and preparation. While the IT team is working on data and your users are engaged in planning and testing processes, your management team should also evaluate issues like these:

- Risk assessment. What are the possible risks of using AI for employees, clients, and other stakeholders? Look at legal, regulatory, ethical, financial, and labor aspects. Think about how changes in your workflows, and in the way people train and do their jobs, might affect the business in the future.

- Contract review. Do any of your contracts require you to transfer rights to your work product? For example, do you create marketing campaigns, custom software, product designs, packaging, training materials, or other deliverables that your client expects to "own" once completed? If so, be especially cautious. Because AI-generated work cannot be copyrighted, you cannot legally transfer copyright in that work. This is another reason why AI should not be used as the final output in situations where intellectual property rights are part of the agreement. At most, AI can be used as a drafting or ideation tool, but the final deliverable must come from human authorship if you are expected to grant copyright or ownership to a client.

Testing AI Results

Testing always matters, and it matters more when AI is involved. Effective testing of an ERP system, for example, requires being able to trace how inventory valuations were determined, following raw material costs, production costs, and labor through each stage until you arrive at the final valuation. These valuations are then tested and confirmed through running accounting reports like trial balances. That level of traceability is essential for accuracy and for explaining results to tax authorities.

AI-inclusive software needs the same treatment. Testing should confirm that you can recreate how the system gathered information, assembled it, and arrived at its recommendations. If you cannot follow the path, then you cannot be sure the result is reliable, and you cannot defend it internally, or to customers, auditors, or regulators.

User Training

The return you get from any software investment depends heavily on how well people are trained. Every implementation should include:

- Training needs assessment. Identify what different groups of users need to learn based on their roles and responsibilities.

- Training plan. Outline the curriculum, schedule, and delivery methods (classroom, online learning, peer training, or blended). Specify who will train, how they will be trained, and what resources they will use.

- Training content. Provide manuals, quick-reference guides, and online resources that are clear, role-specific, and easy to use.

- Training delivery. Make sessions interactive and practical, with plenty of hands-on practice. Respect people's time and attention.

- Training evaluation. Collect feedback and measure whether users can actually apply what they've learned. Adjust the training program as needed until your users are confident and capable.

- Ongoing training. Jobs, software, and processes change. Plan to offer ongoing training in the form of refresher sessions, online modules, and one-to-one support to keep skills current.

Even if your software vendor provides training, it will not be this thorough. Vendors tend to focus on showing how the system works, not on tailoring training to your processes, your culture, or the different needs of your user groups. So get your HR team involved, or look for a good implementation partner that will provide this deeper level of preparation and support. Make sure the training fits your business, and not just the software.

AI Additions to Training

If the software includes AI features, the training program should also cover:

- Overview of AI. Users don't need to be data scientists, but they do need to understand the basics of how AI works. This will help them recognize when the system's recommendations make sense and when something needs to be questioned or escalated.

- Intellectual property (IP). Most employees have little working knowledge of IP, which has not been a major issue until now. As soon as they begin using AI to generate results, they will need clear guidelines about company intellectual property standards and what can and cannot be done. These guidelines should be tailored to their roles. For example, the intellectual property considerations for someone using AI in product lifecycle management will be very different from those for someone using AI to generate marketing copy.

- Bias and fairness. Employees should be trained to spot signs of bias or stereotyping in AI-generated content or recommendations, and know how to report concerns.

- Model updates. Because AI systems evolve, the outputs they generate, such as candidate rankings, pricing suggestions, or marketing recommendations, may change after the vendor updates the model. Users need clear communication about those updates and guidance on what to look for when results start to shift.

Pre-Deployment: Customer Disclosures

Before implementing software changes that affect customers, it is a common practice to let them know what's happening, how it might affect them, and what to do if they encounter problems. For example, a company implementing an ERP system will often forewarn customers about the implementation dates and possible delays. With AI, these disclosures are even more important, because AI systems can blur the line between first-party and third-party data.

- **First-party data** is the information you collect directly: sales records, website analytics, loyalty data, customer surveys, call transcripts, and so on.

- **Third-party data** is collected from sources outside your control.

When your first-party data is fed into an AI tool, and that tool uses your data to improve its own model, your information may in effect become third-party data: data available to the vendor and potentially to other customers using the same system.

For this reason, companies should disclose to customers when AI is being used, explain how customer data may be used in the process, and offer an opt-out where required. Privacy laws in the U.S. are still patchy, but Europe's GDPR is clear on this point, and U.S. law is moving in a similar direction as more states adopt their own privacy frameworks.

Same Implementation Requirements With a New Layer of Responsibility

Technology implementation has always been about planning, testing, and training. AI doesn't change that, but it does expand it. AI adds new questions to ask, new risks to account for, and new training your people will need. Businesses that treat AI as just another "feature" and skip those steps will find themselves exposed. Businesses that treat it as a new layer of responsibility will be better prepared to experience the benefits while avoiding the hazards.

Post Deployment

Business software continuously evolves, so the post-implementation phase of any software implementation involves a thorough project review, ongoing support and maintenance, and continuous improvement. These reviews should include:

- Functional reviews to ensure the software continues to meet the functional requirements originally specified.

- Technical reviews to assess performance (speed, reliability, responsiveness), scalability, and security of the software.

- User reviews to understand the software's usability, accessibility, and the overall experience of the users.

- Quality assurance reviews to assess the overall quality of the software and identify issues or defects that must be addressed.

- Compliance reviews to ensure the software is meeting relevant legal, regulatory, and industry standards.

- Cost-benefit and cost-of-ownership reviews to assess return on investment (ROI) and opportunities for cost savings or improved efficiencies.

○ ○ ○

AI Adds Responsibilities for Ongoing Evaluation

When software includes AI features, post-deployment responsibilities expand:

- Dataset updates. At least once a year, request updates from your supplier about new or expanded datasets used to train the system. Confirm whether these changes alter the intellectual property or privacy assurances that were given at time of purchase.

- Bias assessment. Compare the system's recommendations and results with your benchmarks, goals, and past performance. Look for patterns that suggest unfair treatment, such as certain groups consistently being overlooked, scored lower or offered less favorable terms without a legitimate business reason. It is important to distinguish between intentional segmentation, where you design different offers for different customer groups, and unintended bias, where the system produces unequal results that undermine fairness and trust. Bias detection should be an ongoing practice, not a one-time test.

- Accuracy assessment. Track accuracy and reliability over time. A tool that performed well in the first six months can drift away from acceptable accuracy if the data it's learning from changes.

- Model updates. Ask vendors how often their models are updated, how those updates are communicated, and whether you have a say in whether to accept them. Model updates can change behavior in ways that affect compliance, accuracy, and risk.

Staying Current with the Legal Environment

As we discussed earlier, the legal and regulatory environment around AI is still lagging behind the technology. Companies cannot assume today's practices will be acceptable tomorrow. That means staying vigilant: monitor legal, regulatory, and industry best-practice developments, and be prepared to adapt your policies and systems quickly as new guidelines emerge.

Post-deployment is not a "set it and forget it" stage. Business software keeps changing, and AI software changes even faster. The models are retrained, new features are added, and sometimes the way the system produces its results shifts without much notice. That means the work doesn't stop once you go live.

Regular functional, technical, user, quality, compliance, and financial reviews are how you make sure the tool you bought is still doing the job you need it to do. With AI in the mix, you also have to keep asking: what new data is being used, are the results still accurate, are any groups being treated unfairly, and are we still comfortable with the way this system is operating? These aren't one-time questions. They are questions you'll need to come back to again and again, because the technology itself keeps evolving.

The hidden consequences of an AI system risk being the most expensive. A change in the model could quietly alter how your pricing engine sets discounts, or how your analytics system calculates profitability. A customer-facing AI that once answered questions well could suddenly start making mistakes after an update, and your staff may not notice until customers are already fuming. If you aren't checking in regularly, you may not realize the system has gone off course until the damage is done.

AI can be a powerful partner in operations and decision-making, but only if you manage it actively. Ongoing evaluation will help ensure your software continues to deliver the benefits you paid for, without introducing unacceptable risks. Stay watchful, and you will be in a stronger position to adapt to both technological changes and new rules. Those that assume the system will just run itself may find out the hard way that the real cost of AI isn't in the purchase price, but in the consequences of neglect.

AI Maintenance Checklist:
What to Review Every Year

AI isn't "set it and forget it." Run through these reviews at least once per year:

☐ Dataset changes. Ask your vendor if new data has been added to train the system. Does it affect any privacy or intellectual property assurances you were given?

☐ Accuracy. Check if the system's results are still meeting your expectations. Compare against benchmarks, past performance, or human checks.

☐ Bias. Review whether the system is treating any groups unfairly or creating outcomes that don't align with your diversity and inclusion standards.

☐ Model updates. Find out how often the AI is updated, how those changes are communicated to you, and whether they affect compliance or business processes.

☐ Compliance. Confirm you are still meeting privacy and data protection requirements (GDPR, CPRA, or other state or industry standards).

☐ User experience. Ask employees how well the tool is working in daily use. Have updates made it harder or easier for them to do their jobs?

☐ Return on investment. Look at costs versus benefits. Is the software still paying for itself in efficiency, accuracy, or customer satisfaction?

Treat this as an annual health check for your AI systems. It keeps small problems from turning into big ones and helps you make sure the tools are still working for you, not against you.

Conclusion

History books of the future will identify the broad application of AI as one of the pivotal moments in history. We
are experiencing the introduction of a technology that
changes the very dynamics of competition, customer
expectations, and work itself.

Businesses that opt out of AI-powered software will sacri-
fice opportunity. Soon enough, opting out won't even be
possible; if you are buying software, you are buying AI.
The real choice is in how you will put it to work.

Wise use depends on more than downloading the latest
tool. It requires closing the small business technology
gap: building the digital backbone, skills, and data maturity that let you take advantage of AI instead of being

overwhelmed by it. Without that, AI will only magnify the cracks already in your business.

There's also the workforce paradox. If AI does more of the repetitive work, how will we train the people? Every profession relies on entry-level practice to build judgment and skill. Leaders will have to be intentional about redesigning work so employees still learn, grow, and advance. Efficiency without development is a short-term win, but a long-term loss.

I would be remiss if I didn't also mention the role of policy. Everything we have covered in this book focuses on what you can do inside your own business, but the environment you operate in matters just as much. Recommendations like those from the Goldman Sachs 10,000 Small Businesses Voices community, which focus on digital literacy, access to tools, financing, and workforce training, are not abstractions. They will directly affect how well small and midsize companies can compete in the AI era.

The hardest part of AI is that it pushes us into areas where we don't know what we don't know. That uncertainty is exactly where leadership matters most. AI can help businesses work faster and generate ideas more quickly, and it can take repetitive effort off your team's plate so they have more energy for higher-value work. But the payoff only comes if you keep human oversight, judgment, empowerment, and accountability firmly in place.

Do that, and small businesses will not just keep pace with change, they will be in a position to lead it.

Sound business practices will endure long after the hype around AI has faded. If you stay focused on transparency, fairness, accountability, and alignment with your values, you will be prepared to adapt. That's how you protect your business, your people, and your competitiveness in a world that is being reshaped faster than any of us imagined.

o o o

Downloads & Links

AI Readiness Assessment
Assess your business readiness to implement AI.
werx.me/AI-ready

Business Requirements Document Template
A template with examples to help you get started creating any technology business requirements document.
werx.me/brd

Goldman Sachs 10ksb Voices
Policy recommendations
werx.me/voices

ISO/IEC Explainer
A plain-English explainer document to help professionals without technical comliance backgrounds better understand the ISO/IEC Framework. Includes links and examples.
werx.me/ISO-easy

NIST Explainer
A plain-English explainer document to help professionals without technical comliance backgrounds better understand the NIST framework.
werx.me/nist-easy

NIST Policy
Link to download of NIST AI Risk Management Framework (AI RMF 1.0).
werx.me/nist

NIST Quick Start

Link to download of NIST's quick start document for implementing AI RMF 1.0.

werx.me/nist-start

Glossary

Accountability (in AI)
Clear responsibility for how AI systems are used, including the ability to explain, defend, and own outcomes.

AI (Artificial Intelligence)
Computer systems that perform tasks requiring human-like intelligence, such as recognizing speech, generating text, spotting patterns, or making decisions.

AI RMF (Artificial Intelligence Risk Management Framework)
A guide developed by NIST in 2023 to help organizations use AI responsibly. It is built around four functions: Govern, Map, Measure, and Manage.

Algorithm
A set of rules or instructions a computer follows to process information and produce results.

API (Application Programming Interface)
A set of rules that allows one piece of software to connect with another and exchange information. APIs make it possible for different programs or services to "talk" to each other — for example, when an AI tool plugs into a cloud service to generate text or images. An easy way to think of it: an API is like a waiter in a restaurant, carrying your order to the kitchen (the software system) and bringing the meal (the data or result) back to you.

ATS (Applicant Tracking System)
HR software that scans, ranks, and manages job applicants.

AWS (Amazon Web Services)
A major cloud services platform, often used for hosting data and AI tools.

BCG (Boston Consulting Group)

A global consulting firm, cited in research on AI's role in supply chain and forecasting.

Bias

Systematic errors or unfair outcomes in decision-making, often caused by flawed or incomplete data, or by human assumptions built into systems.

BI (Business Intelligence)

Software that gathers and analyzes business data to support better decision-making.

Black box

A system that produces outputs without showing the internal reasoning that led to them.

Business Requirements Document

A written record that captures what your company needs a new system or project to accomplish, including functions, performance expectations, and outcomes. This document serves as the benchmark for selecting vendors and measuring success.

CCPA (California Consumer Privacy Act)

California's original state privacy law giving residents more control over their data.

Change management

The leadership work of helping people adapt to new systems, processes, or strategies.

Competitive baseline

The current standard of performance or expectation that businesses must meet to remain competitive

Compliance

The practice of following laws, regulations, and contractual requirements that apply to your business.

Computer Vision

A field of AI that enables computers to interpret and act on visual information from the world, such as recognizing faces in photos, detecting defects on a factory line, or reading handwritten text.

CPRA (California Privacy Rights Act)

A 2020 update to the CCPA that expanded consumer rights and created stronger enforcement through the California Privacy Protection Agency.

CRM (Customer Relationship Management)

Software for managing customer relationships, sales pipelines, and marketing campaigns.

Data governance

The rules and processes your company uses to collect, manage, and protect data.

Data Normalization

The process of putting data into a consistent format so it can be compared and analyzed accurately. For example, one system might record dates as "01/05/2025" while another uses "2025-01-05." Normalization makes sure all the dates look the same. In business, it can also mean adjusting values (like sales figures from different regions) so they can be compared fairly.

Data ownership

The rights and responsibilities tied to the information your business creates or holds.

Data proximity problem

The risk that errors or bias in data become harder to detect the farther you get from the original transactions. For example, someone entering a purchase order in an ERP system can usually spot if something looks wrong, but once those orders

are aggregated in a BI dash-
board, mistakes are harder to
notice even though they still
shape decisions.

Dataset

A collection of information
— numbers, text, images, or
other data — used to train or
test an AI system.

DEI (Diversity, Equity, and Inclusion)

Efforts to create fairness,
representation, and belong-
ing in the workplace.

Deep Learning

A type of machine learning
that uses many layers of
interconnected neurons in a
neural network to automatically learn complex patterns
from large amounts of data.

ERP (Enterprise Resource Planning)

Software that integrates core
business processes such as
finance, supply chain, HR,
and inventory.

EU AI Act (European Union Artificial Intelligence Act)

The first comprehensive law
regulating AI, passed in 2024.
It classifies AI by risk and
requires transparency and
accountability for high-risk
systems.

Explainability

The ability to understand
and show how and why an AI
system produced a specific
result.

Feature extraction

The process of pulling out
the most important or useful
pieces of information from a
dataset so an AI system can
focus on what matters. For
example, instead of analyzing
every single pixel in a photo,
the system might extract
features like edges, shapes, or
colors that help it recognize
what's in the image.

o o o

FTC (Federal Trade Commission)
A U.S. government agency that enforces consumer protection and competition laws; increasingly active in AI oversight.

G7 (Group of Seven)
An intergovernmental forum of advanced economies: Canada, France, Germany, Italy, Japan, the United Kingdom, and the United States.

GDPR (General Data Protection Regulation)
The European Union's strict data protection and privacy law, which gives individuals strong rights over their personal data.

Generative AI
AI that can create new content — such as text, images, video, or music — rather than just analyzing existing data.

Governance
The way a company makes and enforces decisions. In the context of AI, it means setting clear rules, responsibilities, and oversight so technology is used responsibly, legally, and in line with the company's values.

Hallucinations (in AI)
When an AI system confidently produces information that sounds correct but is actually false or made up — for example, inventing a source, statistic, or fact that does not exist.

HR (Human Resources)
The part of a business that manages hiring, employee development, performance, and workplace culture.

HBR (Harvard Business Review)
A leading management and business publication, cited in the book.

IEC (International Electrotechnical Commission)

A global standards body for electrical, electronic, and related technologies.

Implementation Partner

A third-party company that helps configure, customize, and roll out software for your business. In ERP systems this role is often called a value-added reseller (VAR), while in areas like CRM or HRIS it is usually referred to as an implementation partner. These partners handle tasks such as data migration, system setup, integrations, user training, and support to make sure the software works effectively in your specific environment.

Intellectual Property (IP)

Legal rights that protect creations of the mind — such as inventions, designs, brand names, written work, or artwork.

ISO (International Organization for Standardization)

An independent global body that develops standards for quality, safety, and accountability.

ISO/IEC 23894

An international standard that provides guidance on managing AI-related risk.

ISO/IEC 42001

An international standard (published 2023) for establishing and managing AI systems responsibly.

KPI (Key Performance Indicator)

A measurable value that shows how well a business is achieving a specific goal.

Labeled Data

A dataset where each piece of information is tagged or classified, serving as an "answer key" for a supervised learning model.

LMS (Learning Management System)

Software for delivering, tracking, and managing training, onboarding, or education programs.

Machine learning

A type of AI where systems learn patterns from data and improve over time without being explicitly programmed.

Minimum standard necessary to compete

The baseline level of performance, capability, or practice that a business must meet in order to remain competitive in its market. Falling below this level doesn't just put you behind — it can make you non-viable compared to peers who are already meeting or exceeding it.

MIT (Massachusetts Institute of Technology)

A U.S. university known for research in science and technology, including AI.

MRP (Manufacturing Resource Planning)

Systems used for production planning, scheduling, and inventory control.

Natural Language Processing

A branch of AI that enables computers to understand, interpret, and generate human language in ways that are useful and meaningful.

NBC (National Broadcasting Company)

A U.S. television network, mentioned in the book as the broadcaster of Chicago Med.

Neural network

A type of AI model inspired by how the human brain works, using layers of digital "neurons" to recognize patterns and make predictions.

Neuron (in neural networks)

A single "node" or computational building block that takes in numbers as input, applies weights and a simple mathematical rule, and passes an output signal to other neurons. When thousands or millions of these neurons are linked in layers, they allow the network to recognize patterns, learn from data, and make decisions.

NIST (National Institute of Standards and Technology)

A U.S. federal agency that develops technology and risk management frameworks, including the AI RMF.

NLP (Natural Language Processing)

A branch of AI that enables computers to analyze, interpret, and generate human language.

Outputs

The results generated by an artificial intelligence (AI) model after it processes an input. These can be in various forms, such as text, images, code, or predictions. The quality and type of an AI's output are directly determined by the training data it was fed and the prompts or instructions it received. For example, when you ask a language model a question, the text it generates in response is the output. When you use an image generator, the image it creates is the output.

Personalization

Tailoring marketing, service, or product recommendations to individual customers based on their behavior or data.

Private dataset

Any dataset that is owned by a specific group or individual and not available to the public.

Proprietary Data

Information that a business owns and controls because it is unique to the company. This can include customer insights, trade secrets, pricing strategies, product formulas, training materials, or other data that provides a competitive advantage. Unlike general operational data (such as invoices or schedules), proprietary data should be carefully protected and rarely shared outside the business.

Public dataset

A dataset that is publicly available for machine learning research and development. Public datasets can include vast amounts of copyrighted and private data, data that the original author or owner did not give permission to include in the dataset.

PwC (PricewaterhouseCoopers)

A global professional services firm, cited in research on predictive maintenance.

Reinforcement Learning

An AI model learns to make decisions by taking actions in an environment and receiving rewards or penalties based on its success.

RPA (Robotic Process Automation)

Software that automates repetitive, rules-based business processes such as data entry or invoice handling.

Robotics

The field of technology that designs and builds machines capable of carrying out tasks in the physical world, often guided by sensors, software, and sometimes AI — from factory robots to autonomous vehicles.

SaaS (Software as a Service)

A way of delivering software over the internet, where you subscribe to use it instead of installing and maintaining it on your own computers. The provider hosts the software, handles updates, security, and maintenance, and you access it through a browser or app. This contrasts with traditional software, which you buy as a one-time license and install locally on your own servers or devices. Examples of SaaS include HubSpot, QuickBooks Online, and Dropbox.

Shadow adoption

When employees start using new tools (often free AI tools) without approval, creating risks around data, security, and compliance.

SHL

A global company that provides talent assessment and psychometric testing tools used by organizations to evaluate candidates' skills, abilities, and leadership potential.

SMB (Small and Midsize Business)

The primary audience for this book. SMBs make up most of the economy but generally don't have the scale or resources of Fortune 500 companies.

SME (Small and Medium Enterprise)

The term used outside the U.S. for small and midsize businesses.

Strict Liability Claim

A legal standard where a business can be held responsible for harm or infringement even if it did not intend to cause the problem and did not know it was happening. In copyright law, for example, you can still be sued for infringement if AI-generated content copies someone else's work, even if you had no idea the system used protected material.

Supervised Learning

An AI model is trained on labeled data, meaning it's given inputs and their correct outputs, so it can learn to make predictions.

Synthetic media

Content created by AI that looks or sounds real, such as deepfake videos, AI-generated images, or cloned voices.

Technology gap

The disadvantage a business faces when it lags behind competitors in adopting or maintaining modern digital systems. Outdated tools, weak cybersecurity, or reliance on manual processes widen this gap over time, making it harder to compete as rivals gain speed, efficiency, and insight from better technology.

Training data

The information used to teach a machine learning model to perform a specific task. It consists of a large dataset of examples that the model analyzes to identify patterns and relationships. The quality and quantity of this data are crucial, as they determine the accuracy and effectiveness of the model's performance. For example, to teach a model to identify cats, you would feed it thousands of images labeled as "cat" or "not a cat." The model learns from these examples, gradually getting better at recognizing a cat on its own.

Training dataset

The data that is used to train a machine learning model.

Transparency

Making clear how decisions are made, what data is used, and how systems operate.

Unlabeled Data

A dataset without tags or classifications, which an unsupervised learning model uses to find its own patterns and relationships.

Unsupervised Learning

An AI model is given unlabeled data and learns to find patterns, structures, and relationships within that data on its own.

Validation dataset

A subset of a training dataset; for example, when training a machine to recognize dogs, the subset of dog pictures, videos, and breed names of dogs in the terrier category. Validation datasets are used to evaluate the performance of the machine learning model during its training.

VAR (Value-Added Reseller)

A company that resells software or hardware and adds its own services or customization.

Vendor Stability Risk

The risk that a software provider may go out of business, get acquired, change direction, or stop supporting its product. When this happens, customers can be left without updates, support, or even access to the system they rely on. Small and midsize businesses should always ask vendors about their financial health, long-term plans, and exit policies to reduce the chance of being stranded with unsupported or abandoned software.

Sources

I've done my best to make this book practical and clear, but the ideas don't stand alone. They come from a wide community of researchers, journalists, and practitioners. Here's where you can find the original work that helped shape the thinking here.

Andersen v. Stability AI Ltd., Midjourney Inc., and DeviantArt Inc. 2023. No. 3:23-cv-00201 (N.D. Cal., filed January 13, 2023).

Asilomar Conference on Signals and Systems. Franklin Declaration. 2010.

Azhar, Azeem. 2021. The Exponential Age: How Accelerating Technology Is Transforming Business, Politics and Society. London: Allen Lane.

Berdinis, Michele. 2023. "The AI Thing No One's Talking About." Bee Blog, March 13, 2023. https://beelinelegal.word-press.com/2023/03/13/the-ai-thing-no-ones-talking-about/.

Boston Consulting Group (BCG). 2018. How Diverse Leadership Teams Boost Innovation. January 23, 2018. https://www.bcg.com/publications/2018/how-diverse-leadership-teams-boost-innovation.

Brynjolfsson, Erik, Ruyu Chen, and Bharat Chandar. 2025. "Canaries in the Coal Mine? Six Facts about the Recent Employment Effects of Artificial Intelligence." Stanford Digital

Economy Lab Working Paper, August 26, 2025. (Primary dataset via ADP payroll data.)

Brynjolfsson, Erik, Danielle Li, and Lindsey R. Raymond. 2023. "Generative AI at Work." NBER Working Paper no. 31161. Cambridge, MA: National Bureau of Economic Research. https:// doi.org/10.3386/w31161.

Deloitte. 2017. Diversity and Inclusion: The Reality Gap. Deloitte University Press, February 2017. https://www2.deloitte. com/insights/us/en/deloitte-review/issue-22/diversity-and-inclusion-at-work-eight-powerful-truths.html

DePillis, Lydia, and Steve Lohr. 2023. "Tinkering With ChatGPT, Workers Wonder: Will This Take My Job?" The New York Times, March 28, 2023. https://www.nytimes. com/2023/03/28/business/economy/jobs-ai-artificial-intelligence-chatgpt.html.

Doshi-Velez, Finale, and J. E. Korteling. 2018. "Accountability of AI under the Law: The Role of Explanation." UC Berkeley Public Law Research Paper (18-10).

European Commission, High-Level Expert Group on Artificial Intelligence. Ethics Guidelines for Trustworthy AI. Brussels: European Commission, April 2019. https://digital-strategy. ec.europa.eu/en/library/ethics-guidelines-trustworthy-ai.

Federal Register. 2023. "Copyright Registration Guidance: Works Containing Material Generated by Artificial Intelligence." Federal Register, March 16, 2023. https://www. federalregister.gov/documents/2023/03/16/2023-05321/

copyright-registration-guidance-works-containing-material-generated-by-artificial-intelligence.

Financial Times. 2024. "The Problem of 'Model Collapse': How a Lack of Human Data Limits AI Progress." Financial Times, July 2024. https://www.ft.com/content/ae507468-7f5b-440b-8512-aea81c6bf4a5.

Financial Times. 2025. "On-the-Job Learning Upended by AI and Hybrid Work." Financial Times, July 2025. https://www.ft.com/content/071089b8-839a-4f96-af79-394c08a146d1.

Future of Life Institute. 2017. Asilomar AI Principles. Asilomar Conference on Beneficial AI, January 2017. https://futureoflife.org/ai-principles.

Future of Life Institute. 2023. "Pause Giant AI Experiments: An Open Letter." March 22, 2023. https://futureoflife.org/open-letter/pause-giant-ai-experiments/.

Gates, Bill. n.d. "Automation Applied to an Inefficient Operation Will Magnify the Inefficiency." QuoteFancy. Accessed August 2025. https://quotefancy.com/quote/775354/Bill-Gates-Automation-applied-to-an-inefficient-operation-will-magnify-the-inefficiency.

Green Software Foundation. 2025. Green AI Position Paper. Green AI Committee, May 13, 2025. https://greensoftware.foundation/articles/green-ai-position-paper.

Hanson, Matthew. n.d. "Registry of Open Data on AWS." Accessed March 26, 2023. https://registry.opendata.aws/.

Harvard Business Review. 2020. "The Secret to AI Is People." August 24, 2020. https://hbr.org/2020/08/the-secret-to-ai-is-people.

Hill, Andrea. 2025. "AI Job Disruption: What It Means for Companies and Competitiveness." Forbes, August 27, 2025. https://www.forbes.com/sites/andreahill/2025/08/27/ai-job-disruption-what-it-means-for-companies-and-competitiveness/.

Hill, Andrea. 2025. "AI Replacing Entry-Level Jobs: The Impact on Workers and the Economy." Forbes, August 27, 2025. https://www.forbes.com/sites/andreahill/2025/08/27/ai-replacing-entry-level-jobs-the-impact-on-workers-and-the-economy/.

Hill, Andrea. 2025. "Artificial Intelligence, AI, Business Growth." LinkedIn, 2025. https://www.linkedin.com/posts/andreahill-consulting_artificialintelligence-ai-businessgrowth-activity-7367184057559904256-_r1s.

Hill, Andrea. 2025. "If AI Does the Work, How Do We Train the People?" AndreaHill.today, August 2025. https://andreahill.today/business-articles/if-ai-does-the-work-how-do-we-train-the-people

Hill, Andrea. 2025. "The Small Business Technology Gap, and How to Bridge It." Forbes, August 12, 2025. https://www.forbes.com/sites/andreahill/2025/08/12/the-small-business-technology-gap-and-how-to-bridge-it.

Hill, Andrea. 2025. "The Strategic Ripple Effect: Preparing for an AI Era We're Already In." LinkedIn Pulse, 2025. https://www.

linkedin.com/pulse/strategic-ripple-effect-preparing-ai-era-were-already-andrea-hill-sdrkc/.

Hill, Andrea. 2025. "Why 95% of AI Pilots Fail, and What Business Leaders Should Do Instead." Forbes, August 21, 2025. https://www.forbes.com/sites/andreahill/2025/08/21/why-95-of-ai-pilots-fail-and-what-business-leaders-should-do-instead/.

IEEE (Institute of Electrical and Electronics Engineers). 2010. Proceedings of the Forty-Fourth Asilomar Conference on Signals, Systems, and Computers (Asilomar 2010), edited by M. Matthews. Monterey, CA: IEEE. proceedings.com.

IEEE (Institute of Electrical and Electronics Engineers). 2016. Ethically Aligned Design: A Vision for Prioritizing Human Well-being with Autonomous and Intelligent Systems. IEEE Global Initiative on Ethics of Autonomous and Intelligent Systems. https://ethicsinaction.ieee.org.

IEEE (Institute of Electrical and Electronics Engineers). Ethically Aligned Design: A Vision for Prioritizing Human Well-being with Autonomous and Intelligent Systems. IEEE Global Initiative on Ethics of Autonomous and Intelligent Systems, 2016. https://ethicsinaction.ieee.org.

Korn Ferry. 2018. Humans and Machines: The Role of Talent in the Future of Work.

McKinsey & Company. 2020. Diversity Wins: How Inclusion Matters. May 2020. https://www.mckinsey.com/capabilities/people-and-organizational-performance/our-insights/diversity-wins-how-inclusion-matters.

MLQ. 2025. State of AI in Business 2025 Report. MLQ Research, v0.1. https://mlq.ai/media/quarterly_decks/v0.1_State_of_AI_in_Business_2025_Report.pdf.

NBER. 2023. "Measuring the Productivity Impact of Generative AI." NBER Digest, June 1, 2023. https://www.nber.org/digest/20236/measuring-productivity-impact-generative-ai.

PwC. 2025. "Using AI to Fast-Track Manufacturing Operations." PwC India, February 3, 2025. https://www.pwc.in/research-and-insights-hub/immersive-outlook-5/using-artificial-intelligence-to-fast-track-manufacturing-operations.html.

SHL. 2018. The Impact of Artificial Intelligence on Talent Assessment.

Thaler v. Perlmutter, 2023 WL 5333236 (D.D.C. Aug. 18, 2023).

United Nations Educational, Scientific and Cultural Organization (UNESCO). 2021. Recommendation on the Ethics of Artificial Intelligence. Paris: UNESCO. November 2021. https://werx.me/unesco.

UNESCO. 2023. "Ethics of Artificial Intelligence." UNESCO, February 3, 2023. https://www.unesco.org/en/artificial-intelligence/recommendation-ethics.

U.S. Copyright Office. 2023. Re: Zarya of the Dawn (Registration No. VAu001480196). Washington, DC: U.S. Copyright Office. February 21, 2023.

Wikimedia Foundation. 2016. "List of Datasets for Machine-Learning Research." **Wikipedia, January 12, 2016.** https://

en.wikipedia.org/wiki/List_of_datasets_for_machine-learning_research.

Wilkins, Joe. 2025. "Company Regrets Replacing All Those Pesky Human Workers With AI, Just Wants Its Humans Back." Futurism, May 13, 2025. https://futurism.com/klarna-openai-humans-ai-back.

World Economic Forum. 2025. Future of Jobs Report 2025: 78 Million New Job Opportunities by 2030 but Urgent Upskilling Needed to Prepare Workforces. Geneva: World Economic Forum, January 7, 2025. https://www.weforum.org/press/2025/01/future-of-jobs-report-2025-78-million-new-job-opportunities-by-2030-but-urgent-upskilling-needed-to-prepare-workforces/.

About the Author

Andrea Hill has spent more than 35 years at the forefront of business growth and transformation, guiding companies to use technology as a driver of strategic results. She is the CEO and Founder of Hill Management Group, LLC, which operates four specialized brands: StrategyWerx, WerxMarketing, MentorWerx, and ProsperWerx. Each brand focuses on a different aspect of helping small and midsize businesses remove barriers to growth and build sustainable success.

Her career includes CEO and senior executive roles in international manufacturing, supply chain, publishing, and marketing firms, where she became known for combining clear strategy with practical execution. She is the creator of the WeRx 360™ Growth Framework, a modular system that helps leaders align strategy, operations, marketing, sales, talent, and technology to scale with confidence. Around the world, Andrea has built a reputation as a "fix-it" CEO, stepping into organizations at pivotal moments to realign strategy, strengthen operations, rebuild cultures, and apply technology in ways that return businesses to health and profitability.

Andrea is the author of *The How-to-Hire Handbook for Small Business Owners*, co-author of several business books, and a regular contributor to Forbes, trade publications, and her widely read blog, andreahill.today. She contributes to national policy as part of the Goldman Sachs 10,000 *Small Businesses Voices* initiative, advocating for innovation and digital adoption to strengthen the small business economy.

In addition to her professional work, Andrea is an avid reader, daily swimmer, and dedicated gardener with a passion for cultivating both outdoor and indoor landscapes. Today she makes her home in Wisconsin, where, surrounded by family, she balances global business leadership with a lifelong curiosity and love of learning.

About Hill Management Group and Our Brands

This book is one expression of the work we do every day: helping business leaders align strategy, people, and technology to build resilient, vibrant companies.

Hill Management Group is built on a rare combination of strengths: seasoned executives with deep expertise in strategy, sales, marketing, operations, product development, HR, and manufacturing, paired with advanced technical fluency in business technology. Our team is not only experienced in running companies, we are also certified across dozens of technology platforms and familiar with dozens more. This technology-agnostic perspective allows us to help clients choose, implement, and integrate the right tools for their businesses without bias.

Where most firms offer either strategic advice *or* technology implementation, we bring both together. That blend of business discipline and technical execution makes us uniquely equipped to help small and midsize companies compete and grow in a technology-driven economy.

With team members across the United States, Canada, the United Kingdom, and Italy, Hill Management Group reflects a global perspective and reach that strengthens our ability to serve clients in diverse markets. Our four specialized brands are each focused on a different aspect of business growth:

StrategyWerx: If you need help making a market shift, adding capacity, introducing new products, implementing technology, or preparing operations for growth, you need StrategyWerx.

WerxMarketing: Sales and marketing systems and services that connect businesses with the customers they are best suited to serve.

MentorWerx: Organizational design, leadership development, and strategic HR to strengthen culture and capacity.

ProsperWerx: Business technology, systems integration, and training that help small and midsize companies operate with the sophistication of larger enterprises.

At the heart of our approach is partnership. We work alongside leaders and teams to ensure that strategy, people, and technology move together, not in silos. The result is growth that is sustainable, cultures that are strong, and systems that serve the business instead of constraining it.

If you'd like to learn more about our approach or explore resources for your business, you're invited to visit us:

www.strategywerx.com

www.werx.marketing

www.mentorwerx.com

www.prosperwerx.com

Or feel free to email Andrea directly, at:

andrea.hill@strategywerx.com